MANAGING
THE NON-PROFIT
ORGANIZATION

Books by Peter F. Drucker

MANAGEMENT

Managing the Non-Profit Organization
The Frontiers of Management
Innovation and Entrepreneurship
The Changing World of the Executive
Managing in Turbulent Times
Management: Tasks, Responsibilities, Practices
Technology, Management and Society
The Effective Executive
Managing for Results
The Practice of Management
Concept of the Corporation
The Daily Drucker with Joseph A. Maciariello
The Essential Drucker

ECONOMICS, POLITICS, SOCIETY

The New Realities
Toward the Next Economics
The Unseen Revolution
Men, Ideas and Politics
The Age of Discontinuity
Landmarks of Tomorrow
America's Next Twenty Years
The New Society
The Future of Industrial Man
The End of Economic Man

FICTION

The Temptation to Do Good
The Last of All Possible Worlds

AUTOBIOGRAPHY

Adventures of a Bystander

MANAGING THE NON-PROFIT ORGANIZATION

Practices and Principles

Peter F. Drucker

*Including interviews with Frances Hesselbein, Max De Pree,
Philip Kotler, Dudley Hafner, Albert Shanker,
Leo Bartel, David Hubbard, Robert Buford,
and Roxanne Spitzer-Lehmann*

Collins Business
An Imprint of HarperCollinsPublishers

The Library of Congress has catalogued the hardcover edition as follows:

Drucker, Peter Ferdinand, 1909–
   Managing the non-profit organization : practices and principles / Peter Drucker. — 1st ed.
     p.   cm.
   Includes index.
   ISBN 0-06-016507-3
   1. Corporations, Nonprofit—Management. 2. Associations, institutions, etc.—Management. I. Title.
HD62.6.D78   1990
658′.048—dc20   89-46525

ISBN-10: 0-06-085114-7 (pbk.)   ISBN-13: 978-0-06-085114-9 (pbk.)

09 10 CC / RRD 10 9 8 7

Contents

PART THREE: Managing for Performance: how to define it; how to measure it

PART FOUR: People and Relationships: your staff, your board, your volunteers, your community

PART FIVE: Developing Yourself: as a person, as an executive, as a leader

Contributors

Frances Hesselbein was from 1976 until 1990 National Executive Director of the world's largest women's organization, the Girl Scouts of the United States of America. She is now President of the Peter F. Drucker Foundation for Non-Profit Management.

Max De Pree is Chairman of Herman Miller, Inc., and of the Hope College Board, and is a member of the board of Fuller Theological Seminary. He is the author of *Leadership Is an Art* (Garden City, N.Y., 1989).

Philip Kotler teaches at the J. L. Kellog Graduate School of Management of Northwestern University in Evanston, Illinois. His pioneering work, *Strategic Marketing for Non-Profit Institutions,* first published in 1971, is now in its fourth edition.

Dudley Hafner is Executive Vice-President and CEO of the American Heart Association.

Albert Shanker is President of the American Federation of Teachers AFL-CIO.

Father Leo Bartel is Vicar for Social Ministry of the Catholic Diocese of Rockford, Illinois.

Reverend David Allan Hubbard is President of Fuller Theological Seminary in Pasadena, California.

Robert Buford is Chairman and CEO of Buford Television, Inc., in Tyler, Texas. He has founded two non-profit institutions, Leadership Network and the Peter F. Drucker Foundation for Non-Profit Management.

Roxanne Spitzer-Lehmann is Corporate Vice-President of St. Joseph Health System, a chain of non-profit hospitals headquartered in Orange, California. She is the author of *Nursing Productivity* (Chicago, 1986).

Preface

Forty years ago, when I first began to work with non-profit institutions, they were generally seen as marginal to an American society dominated by government and big business respectively. In fact, the non-profits themselves by and large shared this view. We then believed that government could and should discharge all major social tasks, and that the role of the non-profits, if any, was to supplement governmental programs or to add special flourishes to them.

Today, we know better. Today, we know that the non-profit institutions are central to American society and are indeed its most distinguishing feature.

We now know that the ability of government to perform social tasks is very limited indeed. But we also know that the non-profits discharge a much bigger job than taking care of specific needs. With every second American adult serving as a volunteer in the non-profit sector and spending at least three hours a week in non-profit work, the non-profits are America's largest "employer." But they also exemplify and fulfill the fundamental American commitment to responsible citizenship in the community. The non-profit sector still represents about the same proportion of America's gross national product—2 to 3 percent—as it did forty years ago. But its meaning has changed profoundly. We now realize that it is central to the quality of life in America, central to citizenship, and indeed carries the values of American society and of the American tradition.

Forty years ago no one talked of "non-profit organizations" or of a "non-profit sector." Hospitals saw themselves as hospitals, churches as churches, Boy Scouts and Girl Scouts as Scouts, and so on. Since then, we have come to use the term "non-profit" for all these institutions. It is a negative term and tells us only what these institutions are not. But at least it shows that we have come to realize that all these institutions, whatever their specific concerns, have something in common.

And we now begin to realize what that "something" is. It is not that these institutions are "non-profit," that is, that they are not businesses. It is also not that they are "non-governmental." It is that they *do* something very different from either business or government. Business supplies, either goods or services. Government controls. A business has discharged its task when the customer buys the product, pays for it, and is satisfied with it. Government has discharged its function when its policies are effective. The "non-profit" institution neither supplies goods or services nor controls. Its "product" is neither a pair of shoes nor an effective regulation. Its product is a *changed human being.* The non-profit institutions are human-change agents. Their "product" is a cured patient, a child that learns, a young man or woman grown into a self-respecting adult; a changed human life altogether.

Forty years ago, "management" was a very bad word in non-profit organizations. It meant "business" to them, and the one thing they were not was a business. Indeed, most of them then believed that they did not need anything that might be called "management." After all, they did not have a "bottom line."

For most Americans, the word "management" still means business management. Indeed, newspaper or television reporters who interview me are always amazed to learn that I am working with non-profit institutions. "What can you do for them?" they ask me, "Help them with fund-raising?" And when I answer, "No, we work together on their mission, their leadership, their management," the reporter usually says, "But that's *business* management, isn't it?"

But the "non-profit" institutions themselves know that they need management all the more because they do not have a conventional "bottom line." They know that they need to learn how to use management as their tool lest they be overwhelmed by it. They know they need management so that they can concentrate on their mission. Indeed, there is a "management boom" going on among the non-profit institutions, large and small.

Yet little that is so far available to the non-profit institutions to help them with their leadership and management has been specifically designed for them. Most of it was originally developed for the needs of business. Little of it pays any attention to the distinct characteristics of the non-profits or to their specific central needs: To their mission, which distinguishes them so sharply from business and government; to what are "results" in non-profit work; to the strategies required to market their services and obtain the money they need to do their job; or to the challenge of introducing innovation and change in institutions that depend on volunteers and therefore cannot command. Even less do the available materials focus on the specific human and organizational realities of non-profit institutions; on the very different role that the board plays in the non-profit institution; on the need to attract volunteers, to develop them, and to manage them for performance; on relationships with a diversity of constituencies; on fund-raising and fund development; or (a very different matter) on the problem of individual burnout, which is so acute in non-profits precisely because the individual commitment to them tends to be so intense.

There is thus a real need among the non-profits for materials that are specifically developed out of their experience and focused on their realities and concerns. It was this need that led a friend of mine, Robert Buford of Tyler, Texas—himself a highly successful business builder—to found Leadership Network, which works on leadership and management in non-profit institutions, and especially in the large pastoral churches, both Protestant and Catholic, that have grown so rapidly in this country in the last twenty years.

I have been privileged to work with Bob Buford from the beginning on this important task and it was out of this experience that the idea for this book emerged. Or rather, what emerged first was a project for a set of audio cassettes designed by me, directed by me, and largely spoken by me on *Leadership and Management in the Non-Profit Institutions* ("The Non-Profit Drucker").

We chose audio cassettes as our first vehicle for two reasons. First, versatility; they can be listened to in one's car driving to work, in one's own home, or at a meeting. But also we thought it important to bring to the non-profit audience the experience and thinking of distinguished people who have built and led important non-profit institutions, both large and small. And this is better done by the spoken word than by a printed text. Accordingly, we produced, in the spring of 1988, a set of twenty-five one-hour audio cassettes. They are being used successfully across the spectrum of non-profit institutions, especially to train new staff people, new board members, and new volunteers.

From the beginning, we also thought of a book that would address itself to the non-profit audience, and a good many of the users of the "Non-Profit Drucker" have urged us to make available the same material in book form. "We want to read you," these cassette users told us, "but in such a way as also to hear the person and especially you, Peter Drucker, as well as the people you interviewed on these tapes."

This book starts out with the realization that the non-profit institution has been America's resounding success in the last forty years. In many ways it is the "growth industry" of America, whether we talk of health-care institutions like the American Heart Association or the American Cancer Society which have given leadership in research on major diseases and in their prevention and treatment; of community services such as the Girl Scouts of the U.S.A. and the Boy Scouts of the U.S.A. which, respectively, are the world's largest women's and men's organizations; of the fast-growing pastoral churches; of the hospital; or of the many other non-profit institutions that have emerged as the center of effective social action in a rapidly changing and turbulent America. The non-profit sector has become America's "Civil Society."

Today, however, the non-profits face very big and different challenges.

The first is to convert donors into contributors. In total amounts, the non-profit organizations in this country collect many times what they did forty years ago when I first worked with them. But it is still the same share of the gross national product (2–3 percent), and I consider it a national disgrace, indeed a real failure, that the affluent, well-educated young people give proportionately less than their so much poorer blue-collar parents used to give. If the health of a sector in the economy is judged by its share of the GNP, the non-profits do not look healthy at all. The share of GNP that goes to leisure has more than doubled in the last forty years; the share that goes to medical care has gone up from 2 percent of the GNP to 11 percent; the share that goes to education, especially to colleges and universities, has tripled. Yet the share that is being given by the American people to the non-profit, human-change agents has not increased at all. We know that we can no longer hope to get money from "donors"; they have to become "contributors." This I consider to be the first task ahead for non-profit institutions.

It is much more than just getting extra money to do vital work. Giving is necessary above all so that the non-profits can discharge the one mission they all have in common: to satisfy the need of the American people for self-realization, for living out our ideals, our beliefs, our best opinion of ourselves. To make contributors out of donors means that the American people can see what they want to see—or should want to see—when each of us looks at himself or herself in the mirror in the morning: someone who as a citizen takes responsibility. Someone who as a neighbor cares.

Then there is the second major challenge for the non-profits: to give community and common purpose. Forty years ago, most Americans already no longer lived in small towns, but they had still grown up in one. They had grown up in a local community. It was a compulsory community and could be quite stifling. Still, it was a community.

Today, the great majority of Americans live in big cities and their suburbs. They have moved away from their moorings, but they still need a community. And it is working as unpaid staff for a non-profit institution that gives people a sense of community, gives purpose, gives direction—whether it is work with the local Girl Scout troop, as a volunteer in the hospital, or as the leader of a Bible circle in the local church. Again and again when I talk to volunteers in non-profits, I ask, "Why are you willing to give all this time when you are already working hard in your paid job?" And again and again I get the same answer, "Because here I know what I am doing. Here I contribute. Here I am a member of a community."

The non-profits *are* the American community. They increasingly give the individual the ability to perform and to achieve. Precisely because volunteers do not have the satisfaction of a paycheck, they have to get more satisfaction out of their contribution. They have to be managed as unpaid staff. But most non-profits still have to learn how to do this. And I hope to show them how—not by preaching, but by giving successful examples.

This book consists of five parts:

    I. THE MISSION COMES FIRST
        —and your role as a leader
   II. FROM MISSION TO PERFORMANCE
        —effective strategies for marketing, innovation, and
          fund development
  III. MANAGING FOR PERFORMANCE
        —how to define it; how to measure it
  IV. PEOPLE AND RELATIONSHIPS
        —your staff, your board, your volunteers, your
          community
   V. DEVELOPING YOURSELF
        —as a person, as an executive, as a leader

In each part I first address the topic. This is then followed by one or two interviews with a distinguished performer in the non-profit field. Each part then concludes with a short, action-focused summary.

I owe a heavy debt to many people. First, I wish to express my thanks to the contributors, the non-profit leaders who so generously gave of their experience and thereby made this book possible. Their achievement in their own institutions shows all of us what can be done and how it should be done.

Then I owe more than I can express in words to my friend Robert Buford, who throughout this entire project has been steadfast in his support, in his advice, in his commitment. His example, that of a successful business leader who is dedicating more and more of his great competence, his time, and his money to leadership in the non-profit, human-change institution, gives guidance to all of us.

Finally, this book owes a great deal to three editors: to Philip Henry, the producer and editor of the audio tapes; to my friend and editor at HarperCollins, Cass Canfield, Jr., who skillfully designed a structure that transforms the spoken into the written word and yet maintains the immediacy of oral communication; and to another old friend, Marion Buhagiar, who, as so often in the past, edited my text with respect both for the integrity of the work itself and for the integrity of the English language.

To all of them, my warmest thanks.

Claremont, California
July 4, 1990

PART ONE

The Mission Comes First

and your role as a leader

1

The Commitment

The non-profit organization exists to bring about a change in individuals and in society. The first thing to talk about is what missions work and what missions don't work, and how to define the mission. For the ultimate test is not the beauty of the mission statement. The ultimate test is right action.

The most common question asked me by non-profit executives is: What are the qualities of a leader? The question seems to assume that leadership is something you can learn in a charm school. But it also assumes that leadership by itself is enough, that it's an end. And that's misleadership. The leader who basically focuses on himself or herself is going to mislead. The three most charismatic leaders in this century inflicted more suffering on the human race than almost any trio in history: Hitler, Stalin, and Mao. What matters is not the leader's charisma. What matters is the leader's mission. Therefore, the first job of the leader is to think through and define the mission of the institution.

SETTING CONCRETE ACTION GOALS

Here is a simple and mundane example—the mission statement of a hospital emergency room: "It's our mission to give assurance to the afflicted." That's simple and clear and direct. Or take the mission of the Girl Scouts of the U.S.A.: to help girls grow into proud, self-confident, and self-respecting young women. There is an Episcopal church on the East Coast which defines its mission as making Jesus the head of this church and its chief executive officer. Or the mission of the Salvation Army, which is to make citizens out of the rejected. Arnold of Rugby, the greatest English educator of the nineteenth century, who created the English public school, defined its mission as making gentlemen out of savages.

My favorite mission definition, however, is not that of a nonprofit institution, but of a business. It's a definition that changed Sears from a near-bankrupt, struggling mail-order house at the beginning of the century into the world's leading retailer within less than ten years: It's our mission to be the informed and responsible buyer—first for the American farmer, and later for the American family altogether.

Almost every hospital I know says, "Our mission is health care." And that's the wrong definition. The hospital does not take care of health; the hospital takes care of illness. You and I take care of health by not smoking, not drinking too much, going to bed early, watching our weight, and so on. The hospital comes in when health care breaks down. An even more serious failing of this mission is that nobody can tell you what action or behavior follows from saying: "Our mission is health care."

A mission statement has to be operational, otherwise it's just good intentions. A mission statement has to focus on what the institution really tries to do and then do it so that everybody in the organization can say, This is *my* contribution to the goal.

Many years ago, I sat down with the administrators of a major hospital to think through the mission statement of the emergency room. It took us a long time to come up with the very simple, and (most people thought) too obvious statement that the emergency room was there to give assurance to the afflicted. To do that well, you have to know what really goes on. And, much to the surprise of the physicians and nurses, it turned out that in a good emergency room, the function is to tell eight out of ten people there is nothing wrong that a good night's sleep won't take care of. You've been shaken up. Or the baby has the flu. All right, it's got convulsions, but there is nothing seriously wrong with the child. The doctors and nurses give assurance.

We worked it out, but it sounded awfully obvious. Yet translating that mission statement into action meant that everybody who comes in is now seen by a qualified person in less than a minute. That is the mission; that is the goal. The rest is implementation. Some people are immediately rushed to intensive care, others get a lot of tests, and yet others are told: "Go back home, go to sleep, take an aspirin, and don't worry. If these things persist, see a physician tomorrow." But the first objective is to see everybody, almost immediately—because that is the only way to give assurance.

The task of the non-profit manager is to try to convert the organization's mission statement into specifics. The mission may be forever—or at least as long as we can foresee. As long as the human race is around, we'll be miserable sinners. As long as the human race is around, there will be sick people. And, as long as the human race is around, there will be alcoholics and drug addicts and the unfortunate. For hundreds of years we've had schools of one kind or another trying to get a little knowledge into seven-year-old boys and girls who would rather be out playing.

But the goal can be short-lived, or it might change drastically because a mission is accomplished. A hundred years ago, one of the great inventions of the late nineteenth century was the tuberculosis sanatorium. That mission has been accomplished, at least in developed countries. We know how to treat TB with antibiotics. And so managers of non-profits also have to build in review, revision, and organized abandonment. The mission is forever and may be divinely ordained; the goals are temporary.

One of our most common mistakes is to make the mission statement into a kind of hero sandwich of good intentions. It has to be simple and clear. As you add new tasks, you deemphasize and get rid of old ones. You can only do so many things. Look at what we are trying to do in our colleges. The mission statement is confused—we are trying to do fifty different things. It won't work, and that's why the fundamentalist colleges attract so many young people. Their mission is very narrow. You and I may quarrel with it and say it's too narrow, but it's clear. It enables the students to understand. And it also enables the faculty to know. And it enables that administration to say, We aren't going to teach accounting.

As you add on, you have to abandon. But you also have to think through which are the few things we can accomplish that will do the most for us, and which are the things that contribute either marginally or are no longer of great significance. A hundred years ago, about the greatest contribution the hospital could make was in obstetrics, though it took a long time before the population accepted that, because childbirth at home in the growing city was perceived to be, well, dangerous, what with infection and untrained people. Well, now I would say that not every hospital should do obstetrics, and a great many don't. Partly because it's become so much safer, so much more predictable. But also because if anything does go wrong, it's so much more critical, so you need a concentration of resources. In a suburban community there just might not be enough volume to do a really good job. So perhaps you don't abandon obstetrics, but you phase it out slowly. On the other hand, fifty or sixty years ago, before the psychotropic drugs, no hospital could do much for mental diseases. Today, almost a majority of people who are mentally sick or endangered can be taken care of in the community hospital, with short-term stays for depression and so on. You can make a major contribution there.

So you constantly look at the state-of-the-art. You look at the opportunities in the community. The hospital isn't going to sell shoes and it's not going into education on a big scale. It's going to take care of the sick. But the specific objective may change. Things that were of primary importance may become secondary or even totally irrelevant. You must watch this constantly, or else very soon you will become a museum piece.

THE THREE "MUSTS" OF A SUCCESSFUL MISSION

Look at strength and performance. Do better what you already do well—if it's the right thing to do. The belief that every institution can do everything is just not true. When you violate the values of an institution, you are likely to do a poor job. In the 1960s, all of us in academia rushed into the urban problem. We were totally incompetent: our values don't fit what are political issues; academicians don't understand power. At the same time, hospitals rushed into what they called health education. Here are the people who come in, such as the diabetic, and before they go home maybe we can teach them how to handle their diet and their stress and so on so that they don't come back. It hasn't worked. That's not what hospitals are good at. Hospitals are *not* good at prevention; hospitals are good at taking care of damage that's already been done.

Look outside at the opportunities, the needs. Where can we, with the limited resources we have—and I don't just mean people and money, but also competence—really make a difference, really set a new standard? One sets the standard by doing something and doing it well. You create a new dimension of performance.

The next thing to look at is what we really believe in. A mission is not, in that sense, impersonal. I have never seen anything being done well unless people were committed.

All of us know the story of the Edsel automobile. Everybody thinks the Edsel failed because Ford didn't do its homework. In fact, it was the best-engineered, the best-researched, the best-everything car. There was only one thing wrong with it: nobody in the Ford Motor Company believed in it. It was contrived. It was designed on the basis of research and not on the basis of commitment. And so when it got into a little trouble, nobody supported the child. I'm not saying it could have been a success. But without that personal commitment, it certainly never could be.

And so one asks first, what are the opportunities, the needs? Then, do they fit us? Are we likely to do a decent job? Are we competent? Do they match our strengths? Do we really believe in this? This is not just true of products, it's true of services.

So, you need three things: opportunities; competence; and commitment. Every mission statement, believe me, has to reflect all three or it will fall down on what is its ultimate goal, its ultimate purpose and final test. It will not mobilize the human resources of the organization for getting the right things done.

2

Leadership Is a
Foul-Weather Job

The most successful leader of this century was Winston Churchill. But for twelve years, from 1928 until Dunkirk in 1940, he was totally on the sidelines, almost discredited—because there was no need for a Churchill. Things were routine or, at any rate, looked routine. When the catastrophe came, thank goodness, he was available. Fortunately or unfortunately, the one predictable thing in any organization is the crisis. That always comes. That's when you *do* depend on the leader.

The most important task of an organization's leader is to anticipate crisis. Perhaps not to avert it, but to anticipate it. To wait until the crisis hits is already abdication. One has to make the organization capable of anticipating the storm, weathering it, and in fact, being ahead of it. That is called innovation, constant renewal. You cannot prevent a major catastrophe, but you can build an organization that is battle-ready, that has high morale, and also has been through a crisis, knows how to behave, trusts itself, and where people trust one another. In military training, the first rule is to instill soldiers with trust in their officers, because without trust they won't fight.

THE PROBLEMS OF SUCCESS

Problems of success have ruined more organizations than has failure, partly because if things go wrong, everybody knows they have to go to work. Success creates its own euphoria. You outrun your resources. And you retire on the job, which may be the most difficult thing to fight. I'm now in California instead of New York University, where I was for twenty years, in part because the Graduate Business School at NYU decided to cut back rather than grow with the growing student demand. That's why I left. When I started to build a management school at Claremont, I made sure that we did not overextend ourselves. I was very careful to ensure that we kept the faculty first rate but small, and that we used adjuncts, part-time people, then built a strong administration. And then we could run with success. If the market grows, you have to grow with it, or you become marginal.

I am arguing these days with our pastor, who wants to keep our church small. This is in a community where we have a lot of young people, students, and a lot of people in retirement homes who want to come to church. My very nice and able pastor likes to keep it small so that he knows everybody. I said to him, "Look, Father Michael, it won't work." Five years after he had come in, the church began to shrink. The lesson for the leaders of non-profits is that one has to grow with success. But one also has to make sure that one doesn't become unable to adjust. Sooner or later, growth slows down and the institution plateaus. Then it has to be able to maintain its momentum, its flexibility, its vitality, and its vision. Otherwise, it becomes frozen.

HARD CHOICES

Non-profit organizations have no "bottom line." They are prone to consider everything they do to be righteous and moral and to serve a cause, so they are not willing to say, if it doesn't produce results then maybe we should direct our resources elsewhere. Non-profit organizations need the discipline of organized abandonment perhaps even more than a business does. They need to face up to critical choices.

Some of these choices are very difficult. I have a friend, a Catholic priest, who is Vicar General of a large diocese. The bishop called him in to deal with the shortage of priests. Which services should they keep and which should they abandon? There is the terrible dilemma of Catholic schools in a big metropolitan archdiocese where 97 percent of the kids are not Catholics and aren't going to be Catholics; they're fleeing the misery of the public schools. I've been arguing with the diocese for years. Some of the priests say, "Our first task is to save souls; it's not to educate people. Let's put our few priests and nuns on our first priority." And I say, "Look, it says in the Bible, 'But the greatest of these is Charity,' and that's what you are doing. You cannot possibly leave those kids in the lurch. That's a value choice, and it's critical that it's faced up to and not pushed under the rug, as we like to do."

Once you acknowledge that, you can then innovate—provided you organize yourself to look for innovation. Non-profit institutions need innovation as much as businesses or governments. And we know how to do it.

The starting point is to recognize that change is not a threat. It's an opportunity. We know where to look for changes.[1] Here are a few examples:

Unexpected Success in Your Own Organization

Some institutions of higher education, for instance, have learned that continuing education of already highly educated adults is not a luxury, or something to bring in additional money, or good public relations. It is becoming the central thrust of our knowledge

[1]See my *Innovation and Entrepreneurship* (Harper & Row, 1985).

society. So, they have organized themselves and their faculties to attract the doctors, engineers, and executives who want and need to go back to school.

Population Changes

About twelve years ago, the Girl Scouts of the U.S.A. realized that demographic shifts in the United States, with the fast growth of minorities, were creating a new frontier for the organization— new needs and the opportunity to change. They now have a 15 percent enrollment of minority kids, which explains why they kept growing even though the total number of girls of scouting age fell quite steadily during that period.

Changes in Mind-Set and Mentality

Very few factors have so altered our view of society as the women's movement of the last twenty years. What opportunities does it create? As you will see a little later on in the interview with Father Leo Bartel in Part Four, it created the opportunity in one diocese to expand dramatically despite a sharp drop in the number of priests and sisters. Another example: about fifteen years ago, one of our largest volunteer organizations, the American Heart Association, realized that, even though its original big job—re-search—was not yet accomplished, a new opportunity had opened to take advantage of the tremendous growth in health awareness by the American public. It decided to redirect its national forces.

The lesson is, Don't wait. Organize yourself for systematic inno-vation. Build the search for opportunities, inside and outside, into your organization. Look for changes as indications of an opportu-nity for innovation. To build all this into your system, you, as the leader of the organization, have to set the example. How can we set up systems to release energy that will allow the proper innovative decisions to be made and implemented and, at the same time, encourage the operation to go on at the necessary level while it is being changed? Let me try to outline a simple series of steps.

First, organize yourself to see the opportunity. If you don't look out the window, you won't see it. What makes this particularly important is that most of our current reporting systems don't reveal opportunities; they report problems. They report the past. Most answer questions we have already asked. So, we have to go beyond our reporting systems. And whenever you need a change, ask: If this were an opportunity for us, what would it be?

Then, to implement the innovation effectively, there are a few points you must be aware of. First, the most common mistake— the one that kills more innovations than anything else—is the attempt to build too much reinsurance into the change, to cover your flank, not to alienate yesterday. The Japanese made that mistake in the one area where their export drive failed significantly: telephones. They had the technology but tried to hedge their bets by selling switchboards that were both electromechanical (and therefore could be plugged into existing old systems) *and* electronic. The electronic switchboards force customers to tear out their old equipment, even though it may be perfectly good. But those who did go either into expansion or improvement of their existing system decided to pull out the old and go straight to the state-of-the-art.

The same sort of mistakes can be found in the pharmaceutical industry and in educational programs. Twenty years ago, a good many hospitals, seeing the trend toward taking care of patients outside the hospital, built outpatient clinics into the hospital. That didn't work. The free-standing surgical clinic, however, *did* work because it was not in the hospital.

Next, you have the problem of organizing the new. It must be organized separately. Babies don't belong in the living room, they belong in the nursery. If you put new ideas into operating units— whether it's a theological seminary or an automobile plant—the solving of the daily crisis will always take precedence over intro-

ducing tomorrow. So, when you try to develop the new within an
existing operation, you are always postponing tomorrow. It must
be set up separately. And yet you have to make sure the existing
operations don't lose the excitement of the new entirely. Otherwise, they become not only hostile but paralyzed.

THE INNOVATIVE STRATEGY

Next, you need an innovative strategy: a way to bring the new
to the marketplace. Successful innovation finds a target of opportunity. Somebody who is receptive, who welcomes the new, who
wants to succeed and, at the same time, has enough stature,
enough clout in the organization so that, if it works for him or her,
the rest of the organization will say, Well, there must be something
to it.

I am always being asked, "If you were running a metropolitan
museum, or a major public library, or a relief or service agency in
a community, would you have part of your organization set up
some kind of small task force committed to R&D or to marketing?
Some group working within the organization that would be weigh-
ing the possibilities of innovation for the organization?"

Well, the answer is yes and the answer is no. Yes, because you
need a few people who do the work, who have the time to do it.
It's hard work. No, because if you isolate the planning, you're
going to end up overlooking perhaps the small but crucial things.
Let me give you a very simple example. The executives of a big
museum decided to move from the old-time museum, which kept
the art works in and the people out, to the modern kind of museum, which is basically an educational community. They set up
a separate planning group, which did a magnificent job planning
exhibitions and publicity and so on. But being isolated from operations, the planners overlooked a few "housekeeping" details. They
forgot, for instance, that you need a much bigger parking lot. Also,
if you suddenly have three hundred fourth graders in, you need

toilets. When they opened, you cannot imagine the pandemonium. And that's typical.

If you first plan and then try to sell, you're going to miss the important things. But you also waste years of time. Selling has to be built into planning, and that means involving the operating people. But don't forget one thing: everything new requires hard work on the part of true believers—and true believers are *not* available part time.

The Churchills may be very rare. But another group is, fortunately, quite common. These are the people who can look at a situation and say: This is not what I was hired to do or what I expected to do, but this is what the job requires—and then roll up their sleeves and go to work. I know a college president who was conned into taking his job with the usual promises by the board that it would raise the money. He came out of tax-supported state universities. He arrived with a wonderful program of faculty recruitment and educational reform, took one good look, and came to me, very unhappy. Somebody has to raise money, he said, otherwise that institution won't be there in five or ten years. And I said, You know, there is only one person who can raise money in a college—the president. And he said, I'm afraid you're right. He found an exceedingly able man on his faculty who for five years ran the school, while the president concentrated on raising money, in which he proved himself incredibly able. He saved that institution.

Let me give you another example of a rural electric cooperative, one of the large ones, founded during the 1930s when the American farmer couldn't get any power. Well, by now everybody has power, so the question is: What do we do now? There was strong sentiment on the board and in the membership for selling out to the nearest large power company. A new chief executive came in, took a look, and said: "Yes, as an electric cooperative we have fulfilled our mission, but as a community development organization, it has only begun. Here is a tremendous farm crisis [this was in the early eighties]. All kinds of basic social services need to be supplied to our farm members, and they can only be supplied by somebody with a distribution system."

He made all the difference. Farm prices are still low and depressed, but this six-county system is one of the few farm areas we have that is, I wouldn't say prosperous, but doing well because of the action this man took seeing the opportunity. And it's not that uncommon. *This* is effective crisis leadership.

HOW TO PICK A LEADER

If I were on a selection committee to choose a leader for a non-profit organization and there were a roster of men and women as candidates, what would I look for? First, I would look at what the individuals have done, what their strengths are. Most selection committees I know are overly concerned with how *poor* the candidate is. Most of the questions I get are not: What is he or she good at, but we think this person is not too good at dealing with students, or what have you. The first thing to look for is strength— you can only perform with strength—and what they have done with it.

Second, I would look at the institution and ask: What is the one immediate key challenge? It may be raising money. It may be rebuilding the morale of the organization. It may be redefining its mission. It may be bringing in new technology. If I looked today for an administrator of a large hospital, I might look for the ability to convert the hospital from a provider of sickness care to a manager of sickness-care providers, because more and more will be done outside the hospital. I would try to match the strengths with the needs.

Then I would look for—call it character or integrity. A leader sets an example, especially a strong leader. He or she is somebody on whom people, especially younger people, in the organization model themselves. Many years ago I learned from a very wise old man, who was head of a large, worldwide organization. I was about twenty, not even that—and he was in his late seventies, famous for putting the right people into the right enterprises all over the globe. I asked him: "What do you look for?" And he said: "I always ask myself, would I want one of my sons to work under that person? If he is successful, then young people will imitate him. Would I want my son to look like this?" This, I think, is the ultimate question.

I've seen lots of businesses and all of us have seen lots of governments survive with mediocre leaders for quite a long time. In the non-profit agency, mediocrity in leadership shows up almost immediately. One difference clearly is that the non-profit has a number of bottom lines—not just one. In business, you can debate whether profit is really an adequate measuring stick; it may not be over the short term, but it is the ultimate one over the long term. In government, in the last analysis, you've got to get reelected. But in non-profit management, there is no such one determinant. You deal with balance, synthesis, a combination of bottom lines for performance.

Certainly, the non-profit executive does not have the luxury of dealing with one dominant constituency, either. In a publicly listed company, the shareholder is the ultimate constituent. In government, it is the voter. When you look at the school board, a public service agency, or a church, however, you have a multiplicity of constituencies—each of which can say no and none of which can say yes. The multiplicity of constituencies is reflected in your boards, your trustees, who are likely to be intensely involved in running the agency. You could say public schools are governmental, but the school board is not governmental. It has the constituency role. That's what causes all the difficulty for school superintendents. They are really public service agencies rather than government agencies.

You can't be satisfied in non-profit organizations with doing adequately as a leader. You have to do exceptionally well, because your agency is committed to a cause. You want people as leaders who take a great view of the agency's functions, people who take their roles seriously—not *themselves* seriously. Anybody in that leadership position who thinks he's a great man or a great woman will kill himself—and the agency.

YOUR PERSONAL LEADERSHIP ROLE

The new leader of a non-profit doesn't have much time to establish himself or herself. Maybe a year. To be effective in that short a time, the role the leader takes has to fit in terms of the mission of the institution and its values. All of us play roles—as parents, as teachers, and as leaders. To work, the role has to fit in three dimensions. First, the role has to fit you—who you are. No comic actor has ever been able to play Hamlet. The role you take also has to fit the task. And, finally, the role has to fit expectations.

One of the more brilliant young men I ever hired as a teacher completely failed in the college classroom. In teaching freshmen, he abdicated his authority, and the kids revolted. He didn't understand that nineteen-year-old freshmen in an undergraduate college expect a teacher to have authority.

You have two things to build on: the quality of the people in the organization, and the new demands you make on them. What those new demands will be can be determined by analysis, or by perception, or by a combination of both. That depends on how you operate. I am a perceptual person. I look. But I've also seen very able and effective people who are totally paper-oriented. They take a sharp pencil and come out right.

There are simply no such things as "leadership traits" or "leadership characteristics." Of course, some people are better leaders than others. By and large, though, we are talking about skills that perhaps cannot be taught but they can be learned by most of us. True, some people genuinely cannot learn the skills. They may not be important to them; or they'd rather be followers. But most of us can learn them.

The leaders who work most effectively, it seems to me, never say "I." And that's not because they have trained themselves not to say "I." They don't *think* "I." They think "we"; they think "team." They understand their job to be to make the team function. They accept the responsibility and don't sidestep it, but "we" gets the credit. There is an identification (very often, quite unconscious) with the task and with the group. This is what creates trust, what enables you to get the task done.

In Shakespeare's *Henry V*, the young prince whose father just died—he's now king—rides out. Falstaff, the old disreputable knight who has been the prince's boon companion in drinking and wenching, calls up to his "Sweet Prince Hal," and the new king rides by without even a look at him. Falstaff is cruelly hurt. He raised the prince because the old king was a very poor father and a cold one, and the young man found warmth only with that disreputable drunkard. Yet Henry is now king and has to set different standards for himself because he is visible. As a leader, you are visible; incredibly visible. And you have expectations to fulfill.

Then there is the story of the one leading German statesman before World War I who saw the catastrophe Europe was sliding into and tried desperately to reverse the trend. He was the ambassador to London in the early days of the century—a leading dove. But he resigned his ambassadorship because the new English king, Edward VII, was a notorious womanizer who liked the diplomatic corps to give him stag parties at which the most popular London courtesans would pop naked out of cakes. The ambassador said he was not willing to be a pimp when he saw himself in a mirror shaving in the morning. I don't think he could have averted World War I. Still, politically, he may have made the wrong decision. And yet, I think, it was the essence of leadership. You are visible; you'd better realize that you are constantly on trial. The rule is: I don't want to see a pimp in the mirror when I shave in the morning. If you do see one, then your people will see one, too.

"To every leader there is a season." There is profundity in that statement, but it's not quite that simple. Winston Churchill in ordinary, peaceful, normal times would not have been very effective. He needed the challenge. Probably the same is true of Franklin D. Roosevelt, who was basically a lazy man. I don't think that FDR would have been a good president in the 1920s. His adrenalin wouldn't have produced. On the other hand, there are people who are very good when things are pretty routine, but who can't take the stress of an emergency. Most organizations need somebody who can lead regardless of the weather. What matters is that he or she works on the basic competences.

As the first such basic competence, I would put the willingness, ability, and self-discipline to listen. Listening is not a skill; it's a discipline. Anybody can do it. All you have to do is keep your mouth shut. The second essential competence is the willingness to communicate, to make yourself understood. That requires infinite patience. We never outgrow age three in that respect. You have to tell us again and again and again. And demonstrate what you mean. The next important competence is not to alibi yourself. Say: "This doesn't work as well as it should. Let's take it back and re-engineer it." We either do things to perfection, or we don't do them. We don't do things to get by. Working that way creates pride in the organization.

The last basic competence is the willingness to realize how unimportant you are compared to the task. Leaders need objectivity, a certain detachment. They subordinate themselves to the task, but don't identify themselves with the task. The task remains both bigger than they are, and different. The worst thing you can say about a leader is that on the day he or she left, the organization collapsed. When that happens, it means the so-called leader had sucked the place dry. He or she hasn't built. They may have been effective operators, but they have not created vision. Louis XIV was supposed to have said, *"L'état, c'est moi!"* (The state, that's me!). He died in the early eighteenth century and the long, not-so-slow slide into the French Revolution immediately began.

When effective non-profit leaders have the capacity to maintain their personality and individuality, even though they are totally dedicated, the task will go on after them. They also have a human existence outside the task. Otherwise they do things for personal aggrandizement, in the belief that this furthers the cause. They become self-centered and vain. And above all, they become jealous. One of the great strengths of Churchill and one of the great weaknesses of FDR was that Churchill, to the very end, when he was in his nineties, pushed and furthered young politicians. That is a hallmark of the truly effective leader, who doesn't feel threatened by strength. In his last years, FDR systematically cut down everybody who showed any signs of independence.

I would not want any person to give his or her life to an organization. One gives one's very best efforts. What attracts people to an organization are high standards, because high standards create self-respect and pride. Most of us want to contribute. When you look at schools where kids learn and schools where kids don't, it's not the quality of the teaching that's different. The school in which kids learn expects them to learn. Many years ago, I did a survey of Boy Scout Councils with tremendous differences in performance. In the performing ones, they expected the volunteers, the scoutmasters, and so on, to put in very hard work. And I mean hard work, not just appearing Friday night for a couple of hours. The ones with high demands attracted the volunteers and attracted and kept the boys. So it is the job of the leaders to set high standards on one condition—that they be performance-focused.

Most leaders I've seen were neither born nor made. They were self-made. We need far too many leaders to depend only on the naturals. The best example of one who surely was not born a leader, had no training, and made himself into a very effective one, was Harry Truman. When Truman became president, he was totally unprepared. An ordinary politician, he was chosen as vice president because he presented no threat to FDR. Truman not only said, "I am president now and the buck stops here," but he also asked, "What are the key tasks?" His entire preparation had been in domestic affairs. He forced himself to accept the fact that the key tasks for his administration were outside the United States and not the New Deal (much to the disappointment of the New Deal liberals, beginning with Mrs. Roosevelt). He forced himself to take a cram course in foreign affairs and to focus—painfully—on what he considered to be key tasks.

In a way, the hospital as we know it today is the creation of a totally obscure and forgotten Catholic hospital administrator of the 1930s and 1940s (who taught me all I know), Sister Justina in Evanston, Indiana. She was the first person to think through what patient care is. For her contributions she got very few thanks in her life, especially not from the physicians, but she was a born leader. She was retiring, shy, understated, very conscious of the fact that her formal education had stopped in first grade in an Irish country school. But there was a job to be done. And that, again and again, is what really makes the leaders. They are self-made.

Douglas MacArthur was a brilliant man and probably the last great strategist, but that wasn't his great strength. He built a team second to none because he put the task first. He was also unbelievably vain, with a tremendous contempt for humanity, because he was certain that no one came close to him in intelligence. Nevertheless, he forced himself in every single staff conference to start the presentation with the most junior officer. He did not allow anybody to interrupt. This contributed incredibly to his ability to build an organization that was willing to fight against the vastly superior enemy and win. It is very clear from his letters that this didn't come easily to him, never. He always had to force himself. It wasn't his nature, but it *was* the key task, and so it had to be done.

Tom Watson, Sr., the creator of IBM, began as a self-centered, imperious man—vain, with a very short fuse. He forced himself to build a team, a winning team. He once let somebody go who I thought was very able and I asked why. Watson told me: "He is not willing to educate me. I am not a technical man, I am a salesman. But this is a technical company, and if they don't educate me in technology, I can't give them the leadership they need." It's that willingness to make yourself competent in the task that's needed that creates leaders.

When Ted Houser took over in the early 1950s, Sears, Roebuck had had twenty-five years of unbroken success. Houser had been a buying strategist and a statistician, purely a figures man. He looked at the company and asked: What does it need so that it can be successful *another* twenty-five years? He concluded that it needed managers. So he forced himself into taking the leadership of Sears' manager development in a very effective and yet very quiet way. Everybody down to the manager of the smallest store knew that the chairman in Chicago was watching him, and would know whether he was developing people. Sears hasn't had a new idea since 1950, yet it remained very successful for twenty-five or thirty years, almost up to 1980, because it had the people. That's what Ted Houser built.

THE BALANCE DECISION

One of the key tasks of the leader is to balance up the long range and the short range, the big picture and the pesky little details. You are always paddling a canoe with two outriggers—balancing—while managing a non-profit. One is the balance between seeing only the big picture and forgetting the individual person who sits there—one lonely young man in need of help. I've heard of hospitals that talk health-care statistics and forget the mother with a crying baby in the emergency room. That kind of failing is fairly easy to correct. Being on the firing line a few days, a few weeks, a year, usually does it. The opposite danger is becoming the prisoner of operations. That's much harder to avoid. The effective people do it very largely through their work in associations and other organizations. The successful chief executive of one of our major community service organizations, one of the very large Scout Councils, sits on three boards of which only one is a community service organization—quite intentionally. And she also sits on an advisory committee of the city government. That way she is forced to see the same issues she faces in her own organization through the other end of the telescope. That works.

I've also seen it done on a smaller, much smaller, scale. A dean I worked with for many years, whom I considered singularly successful, went on the American Council of Deans. I said to him, "Paul, you are so busy, why do you do it?" And he said, "I'm too close to the details. Once a month, I need to see what the overall issues really are." That, too, is a fairly effective way.

Let me say there are always balancing problems in managing non-profits. This is only one example. Another, which I think is even harder to handle, is the balance between concentrating resources on one goal and enough diversification. If you concentrate, you will get maximum results. But it's also very risky. Not only may you have chosen the wrong concentration, but—in military terms—you leave your flanks totally uncovered. And there's not enough playfulness; it doesn't stir the imagination. You need that, so that there will be diversity, especially as any single task eventually becomes obsolete. But diversity can easily degenerate into splintering.

The even more critical balance, and the toughest to handle, is between being too cautious and being rash. Finally, there is timing—and this is always of the essence. You know the people who always expect results too soon and pull up the radishes to see whether they've set root, and the ones who never pull up the radishes because they're sure they're never ripe enough. Those are, in philosophical terms, Aristotelian Prudences, so to speak. How to find the right Mean.

It's actually fairly easy to deal with people who want results too soon. I'm one of them. And I've taught myself that if I expect something to happen in three months, I say, make it five. But I've also seen people who say three years when they should say three months. That's very hard to counteract. As in all Aristotelian means, the first law is "Know thyself." Know what is your degenerative tendency.

I've seen more institutions damaged by too much caution than by rashness, though I've seen both. Maybe I'm conscious of it because I was over-cautious when I ran institutions, or was part of the running. I did not take risks, especially financial risks, I should have taken. On the other hand, I've seen one of the country's universities almost ruined—Pittsburgh, in the 1950s—by a brilliant man who came in and tried to convert what was a fair metropolitan university into a world-class research institution in three years. He thought money would do it. Instead, he almost killed the university, and it has never quite recovered. I've seen the same thing in a museum and the same thing in a symphony orchestra. So, one has to have balance, and again the only advice I can give is to make sure you know your degenerative tendency and try to counteract it.

Then there is the balance decision between opportunity and risk. One asks first: is the decision reversible? If it is, one usually can take even considerable risks. In the non-profit institution, you constantly must gauge whether the financial dimension of a risk is too great. That's all I can say. One looks at the decision: Is it reversible? And what kind of risk is it? Then one asks: Is it a risk we can afford? All right, if it goes wrong, it hurts a little. Or is it a risk that, if things go wrong, will kill us? Or the trickiest of them all, the risk we can't afford not to take. I've been in a similar situation recently. I sit on a museum board—and a big collection was offered to us, way beyond our means. I said, Damn the torpedoes, let's buy it. It's the last chance we have. It'll make us a world-class museum. We'll get the money somehow. The balance decisions are what we need non profit leaders for, whether they are paid or volunteer.

THE DON'T'S OF LEADERSHIP

Finally, there are a few major don't's for leaders. Far too many leaders believe that what they do and why they do it must be obvious to everyone in the organization. It never is. Far too many believe that when they announce things, everyone understands. No one does, as a rule. Yet very often one can't bring in people before the decision; there just isn't enough time for discussion or participation. Effective leaders have to spend a little time on making themselves understood. They sit down with their people and say: This is what we were faced with. These are the alternatives we saw, the alternatives we considered. They ask: What is your opinion? Otherwise the organization will say: Don't these dummies at the top know anything? What's going on here? Why haven't they considered this or that? But if you can say, Yes, we considered it, but still reached this decision, people will understand and will go along. They may say we wouldn't have decided that way, but at least upstairs, they just didn't shoot from the hip.

And the second don't. Don't be afraid of strengths in your organization. This is the besetting sin of people who run organizations. Of course, able people are ambitious. But you run far less risk of having able people around who want to push you out than you risk by being served by mediocrity. And finally, don't pick your successor alone. We tend to pick people who remind us of ourselves when we were twenty years younger. First, this is pure delusion. Second, you end up with carbon copies, and carbon copies are weak. The old rule both in military organizations and in the Catholic Church is that leaders don't pick their own successors. They're consulted, but they don't make the decision. I've seen many cases in business—but even more in non-profit institutions—where able people picked a good number two to succeed them. Somebody who is very able—provided you tell him or her what to do. It doesn't work. Partly out of emotional commitment, partly out of habit, the perfect number two is put into the top spot, and the whole organization suffers. The last time I saw this was in one of the world's largest community chests. Fortunately the number two who was picked by his predecessor because he was so much like her realized after a year that he didn't belong in the top job and was utterly miserable in it—and he left before either he or the organization had been badly damaged. But that is a rare exception. The last don't's are: Don't hog the credit, and Don't knock your subordinates. One of the very ablest men I've seen do this headed one of the most challenging new tasks in a non-profit organization I know. His alumni now work for everybody else but for his organization because the moment they went to work for him, he saw nothing but their weaknesses. He didn't promote any of his people and never sang their praises. A leader has responsibility to his subordinates, to his associates.

Those are the don't's.

The most important do, I have said again and again already: Keep your eye on the task, not on yourself. The task matters, and you are a servant.

3

Setting New Goals

Interview with Frances Hesselbein *

PETER DRUCKER: Frances, of all the new programs you have successfully introduced into your 335 Girl Scout Councils around the country in thirteen years as National Executive Director, which is the one closest to your heart?

FRANCES HESSELBEIN: I would have to say the Daisy Scouts. This is our newest program for little girls, five years old or in kindergarten. In partnership with our Girl Scout Councils, we studied the needs of girls and we studied the American family in all of its configurations, and concluded that girls who are five years old are quite ready for a program working in a small group with two sensitive leaders. In this country today 85 percent of all children five years old are in school all or part of the day.

PETER DRUCKER: That was quite a departure, wasn't it, from the Girl Scout tradition?

FRANCES HESSELBEIN: Yes. Previously, we served girls from seven through seventeen. We moved the Brownie age level back to six because it was very clear as we studied the needs of girls that Brownies were ready at six. It became equally clear that young

*Frances Hesselbein was from 1976 until 1990 National Executive Director of the world's largest women's organization, the Girl Scouts of the United States of America. She is now president of the Peter F. Drucker Foundation for Non-Profit Management.

girls of five were ready for a Girl Scout program designed just for them.

PETER DRUCKER: Were your Councils enthusiastic about the change?

FRANCES HESSELBEIN: I'm afraid that only 70 of the 335 were enthusiastic, wanted to move right then. We had another thirty in the wings thinking more positively about it. But we began with one third of our Councils on board.

PETER DRUCKER: Am I right that you can't order the Councils to do anything?

FRANCES HESSELBEIN: The Councils are chartered, and they have their own volunteer board of directors. They work to meet the special needs of girls in their own areas. So in this case, they really had a choice—move with us or stand back and wait.

PETER DRUCKER: Quite a few of your Councils were, to say the least, dubious, am I right?

FRANCES HESSELBEIN: Yes, they were. But when we were ready to move with the training of the trainers and leaders of Daisy Girl Scouts, we had almost two hundred Girl Scout Councils ready, enthusiastic, and able to open their doors to these newest members.

PETER DRUCKER: How long did it take to go from seventy to two hundred Councils?

FRANCES HESSELBEIN: That took about six months. Within a year the Daisy Scouts were established as one of our most successful endeavors. Three years later, the Daisy Scouts were everywhere in this country. Councils discovered they can offer leadership positions to young women and older women who were reluctant to work with teenagers but who find working with five-year-olds an adventure.

PETER DRUCKER: How many Daisy Scouts do you have now?

FRANCES HESSELBEIN: Approximately 150,000—and growing fast.

PETER DRUCKER: So let me try to play back what I think you told me. First, you were market-driven. You went out and looked at the needs, the wants of the community you serve, and they had changed since you first started seventy-five years ago. So you developed this service that was market-driven. Next, you have to market, you have to persuade, you have to create customers for the new mission because 335 Councils don't have to take a program just because you in New York say so. And the next thing I think you told us is that to make the change, you looked for what I call targets of opportunity—the Councils who really wanted this and were ready to go to work. You didn't worry about the Councils that were non-believers.

FRANCES HESSELBEIN: We began with those Councils ready and eager to move ahead with a new program for five-year-olds. Those not on the sled could wait. We made it very clear that they had a choice. But we were firm in our determination to move ahead with those who were ready and enthusiastic.

PETER DRUCKER: What about those who were ready but not competent?

FRANCES HESSELBEIN: Everyone who wanted to begin the program had to take the new training for trainers and for leaders. We never begin this kind of program without the adult education those women and men needed.

PETER DRUCKER: You said something terribly important. I've seen so many first-rate non-profit services fail because they were just offered, instead of the non-profits' managers making sure that everybody who has to do something knows what has to be done, is trained to do it, has the tools. Did you give your Councils the tools to bring in the new volunteers for this new program?

FRANCES HESSELBEIN: Yes. We created a wonderful handbook for Daisy Scout leaders. We made clear that there should be six to eight girls and at least two leaders in each group. The program had to be educationally sound. Then it had to be carried out in a way that was supportive and helpful. And we have been stressing that the leadership should come from the widest spectrum, not just the mother of the young girl, but young business and professional women. And moving to the other end, older Americans, retired, with lots of energy and interest and willingness to help. I believe this has brought the kind of success, the kind of inclusiveness that is necessary if you're building a large volunteer workforce.

PETER DRUCKER: So you basically spent as much time on thinking through what the program has to be to attract volunteers as you spent on making the program fit the five-year-olds?

FRANCES HESSELBEIN: Yes. Not only the recruitment and placement of volunteers, but designing the training for them to meet their very special needs so that as they moved into work with their group of Daisy Scouts, they could feel very secure.

PETER DRUCKER: How much training does that mean?

FRANCES HESSELBEIN: It would depend upon the person. The staff and volunteers working with potential leaders are very sensitive about their readiness. Training is designed especially for them.

PETER DRUCKER: Now let's switch to another of your successful programs. You have been able to increase the number of volunteers at a time when your traditional volunteer has—I wouldn't say disappeared, but become mighty scarce because many young women no longer sit at home waiting for husbands to come home from work.

FRANCES HESSELBEIN: As we looked at the large core of volunteers, women and men, we realized that they deserved and required superior learning opportunities. Peter, do you remember how we brought the Volunteer Presidents of Girl Scout Councils to California where you gave a seminar on non-profit management? On the other coast we had a team of Harvard Business School faculty give a similar course to the executive directors of our Girl Scout Councils. The quality of those opportunities said something to the volunteers about how this organization needed and respected them and their potential and gifts.

PETER DRUCKER: But how did you get those potential volunteers in the first place?

FRANCES HESSELBEIN: You can't recruit local volunteers from an office in New York. It has to be people in the community who really believe in the mission, who really care about girls, who go out and talk person-to-person with potential volunteers. Our 335 Girl Scout Councils have done a superb job of this.

PETER DRUCKER: Let me try to convert this into general ideas, concepts, and rules. You look at the volunteers as your most important market simply because the number of volunteers you can bring in determines how many girls you can serve. And you make a determined, continued effort to find the right people. Then you treat them, not as volunteers but as unpaid members of the organization. You determine their job, you set the standard, you provide the training, and you basically set their sights high.

That, in my experience, is the secret to the crucial marketing problem of so many non-profit organizations—the volunteer professionals who get their satisfaction out of their work, not the paycheck.

FRANCES HESSELBEIN: You forgot one point—the recognition. It is important that someone says: "Thank you very much, you've made a major contribution." And this too is an important part of the support and care of that volunteer workforce.

PETER DRUCKER: Would the same approach, the same principles, apply to work in the minority communities, where you are more successful than, I would say, any other community service organization in this country?

FRANCES HESSELBEIN: One of the priorities of our National Board of Directors and Girl Scout Councils has been, and is, offering equal access to membership to every girl in the United States. It is important as we reach out to girls in every racial and ethnic group to understand the very special needs, the culture, the readiness of each group. We know that we must find leaders there, whether it is a community of newly arrived Vietnamese or an older, established black community.

PETER DRUCKER: When you took over, the minority membership was small, wasn't it?

FRANCES HESSELBEIN: It was certainly small. The change required daily hard work. It's not enough to have a campaign zoom into a minority community, recruit people, and leave. It requires the most thoughtful kind of planning and including those community leaders in that planning.

PETER DRUCKER: Well, give me an example.

FRANCES HESSELBEIN: In a housing project there are hundreds of young girls, really needing this kind of program, families wanting something better for their children. We work with clergymen, perhaps, with the director of that housing project, with parents—a group of people from that particular community. We recruit leaders, train them right there. In our recruitment brochure we have to demonstrate our respect for that community, our interest in it. Parents have to know that it will be a positive experience for their daughters.

PETER DRUCKER: But what makes you go to that housing project or to that Vietnamese community in the first place?

FRANCES HESSELBEIN: We look at the projections and understand that by the year 2000, one third of this country will be members of minority groups. We have the most remarkable opportunity to serve in new ways. We have to understand what this means to a local Girl Scout Council with many changing ethnic and racial groups within its jurisdiction. To really give leadership to this and to be ready for the year 2000, we developed a national center for innovation. We have a highly skilled staff moving in first to Southern California, where the change is coming so rapidly, working with a small group of Southern Californian Councils, developing models of how a council reaches out to all the girls in its council boundaries and how we really provide that equal access which is so essential.

PETER DRUCKER: Those seven California Councils are already about 30 percent minority populations, right? So that you are actually working on the target of opportunity. They know they need the help. And you also demonstrate effect. If it works here, it's going to work in Buffalo.

FRANCES HESSELBEIN: We chose California. It is the bellwether state, in our opinion, and these models then can be adapted wherever Councils in this country are faced with the opportunity of serving diverse and rapidly changing populations. Theory is not enough.

In 1912, our founder said, "I have something for all the girls."

We take this very seriously. Many people are very apprehensive about the future and what this new racial and ethnic composition will mean to our country. We see it as an unprecedented opportunity to reach all girls with a program that will help them in their growing-up years, which are more difficult than ever before.

PETER DRUCKER: Frances, isn't it pretty typical of the non-profit organization that it has more than one customer? You, for instance, have the girls, but also the volunteers.

FRANCES HESSELBEIN: I believe it is typical. Rarely does a non-profit organization have "a" customer. If we market to only one of our customers, I think we fail.

PETER DRUCKER: And what would your general conclusions be about introducing a new program?

FRANCES HESSELBEIN: You must carefully construct a marketing plan. Not just disseminate information about it, but understand all the ways there are to reach people and use them. Distributing written materials isn't enough. You need people in the marketing chain. And there has to be continuing evaluation—getting feedback on how we are doing. And if a strategy is not working, regroup and move ahead in a different way.

4

What the Leader Owes

*Interview with Max De Pree**

PETER DRUCKER: Max, you have the reputation in your company, but also in the institutions which you serve as a board member, of being the leader in developing people. Is there any one thing you would stress about that?

MAX DE PREE: I would have to begin with a very personal observation, which is that I believe, first of all, that each of us is made in the image of God. That we come to life with a tremendous diversity of gifts. I think from there, a leader needs to see himself in a position of indebtedness. Leaders are given the gift of leadership by those who choose or agree to follow. We're basically a volunteer nation. I think this means that people choose a leader to a great extent on the basis of what they believe that leader can contribute to the person's ability to achieve his or her goals in life. This puts the leader in the position of being indebted—in the sense of what he or she owes to the organization.

One relatively straightforward way of looking at it is that the leader owes certain assets to an organization. In some organizations, that would be the ability to recruit the right people. Another important asset is the ability to raise the necessary funds. Another area isn't quite as clear, and I would put that under the general

*Max De Pree is chairman of Herman Miller, Inc., and of the Hope College Board, and is a member of the board of Fuller Theological Seminary. He is the author of *Leadership Is an Art* (Garden City, N.Y., 1989).

heading of a legacy: the values of the organization. The leader may not be the author of those values, but the leader is accountable for expressing them, making them clear, and ensuring to the people in the organization that the values will be lived up to in a way in which decisions are made. Vision comes under the heading of legacy. Agreed-upon work processes come under this heading. If leaders say, "If you come to work in this organization, I can promise you that we're going to have a participatory process," they are indebted to provide that. An element that is clearly, to me at least, common ground—whether one is in the profit-making or non-profit organization—is that this whole matter of people development needs to be oriented primarily toward the *person,* and not primarily toward the organization.

PETER DRUCKER: You develop people, not jobs, is what you are saying.

MAX DE PREE: Yes, and I'm saying too that when you take the risk of developing people, the odds are very good that the organization will get what it needs.

PETER DRUCKER: But you are also implying, I take it, that you can only develop what the person has. Not what the person ain't got?

MAX DE PREE: That's right. We're talking about building on what people are—not about changing them. To understand their gifts, to understand what their potential is. We tend, in organizations, to put a lot of emphasis on the achievement of goals, but when we're talking about the development of a person, we have a much higher aim. Here we're talking about potential.

    That attitude about people development, I believe, also applies, by the way, to the development of an organization. I think if we focus on goal achievement, we miss the chance we have of realizing our potential. Goal achievement is an annual matter related to the annual plan. But the realization of our potential, that's a life matter.

PETER DRUCKER: Don't you really look at two aspects? You look at the gifts of people, their potential, their strength, what they could be if only they used a little better what they have. But you also look at the objective needs, the objective requirements, the opportunities for accomplishment. Don't you always look inside and out?

MAX DE PREE: You need to make a connection between this matter of realizing potential and doing it in a very real environment. One needs to be accountable, and the accountability needs to be connected to the needs of the organizations.

PETER DRUCKER: And don't you also need achievement to grow?

MAX DE PREE: You do need achievement, and I also happen to feel that that's one of the things for which the leader is partially accountable. I believe the leader needs to assign opportunities and assign work that can be realized. I don't think leaders ought to assign work that's impossible.

PETER DRUCKER: For this person?

MAX DE PREE: For this person, that's right.

PETER DRUCKER: So the leader starts out with what this person really is good at, and then tries to place the person where the strength can redound to performance?

MAX DE PREE: Yes, and, of course, any time we talk about accountability and about achievement, it has to be clear that we are going to delegate thoroughly. Delegate with a certain abandon so that people have space in which to realize potential, in which to be accountable, in which to achieve. I don't believe we can achieve organizational goals without that congruency. I believe it is more the responsibility of the leader to forge that integration than it is of the individual. It's the kind of thing that a follower has a right to expect from a leader.

PETER DRUCKER: You implied a little earlier, Max, that the first duty of a leader was to have followers. In fact, the definition of a leader, the only definition, is somebody who has followers. What is necessary for this? A clear mission? A clear vision?

MAX DE PREE: A leader must have vision. It is natural for a leader to be a person who is primarily future-oriented. I don't mean that to be a duplication of having vision. Those are not exactly the same things. To talk more specifically about the duties of a leader, I happen to believe that the first duty of a leader is to define reality. Every organization, in order to be healthy, to have renewal processes, to survive, has to be in touch with reality.

PETER DRUCKER: How would you define reality for a liberal arts college with its 2,500 students?

MAX DE PREE: One reality, for instance, might be that it happens to be a tuition-driven college. If you don't understand that clearly, you will not put the right amount of emphasis on the recruitment of students. So it's important that the leader sees and defines clearly for the group what reality is.

PETER DRUCKER: A little earlier you said something very important which, in my opinion, very few people in the non-profit institutions yet realize. Most of us still operate on the assumption that people have no choice. They have to take a job. This was true a hundred years ago. But today there are fifty different ways we can earn a living. You call it electiveness, Max, I believe. We have to deserve the person who works for us. We owe him or her, which is what you meant by indebtedness. Because they are not committed to us by necessity; they are committed to us by choice.

MAX DE PREE: People have a lot of choice in where they're going to work, what kind of work they're going to do. They have a lot of choice about mid-career changes. We're only about a generation away from people who, once they had chosen a career, had to stick with it. That's all different today.

PETER DRUCKER: That has to be built into the development processes, I suspect.

MAX DE PREE: Yes. And I think it is related to the kinds of promises that the leader gives. At the heart of that is the whole matter of opportunity. Opportunity is clearly one of the most important things that we seek today in our working lives.

PETER DRUCKER: Opportunity for what?

MAX DE PREE: For self-realization, for being part of a social body that is attractive and rewarding. Opportunity for doing work which will help me to reach my potential. Opportunity to be involved with something that's meaningful. Opportunity to be an integral part of something. We do not develop vital surviving organizations unless we take into account these needs for meaningful work, for a chance at reaching our potential for good social relationships.

PETER DRUCKER: Instead of bemoaning that young people are lazy or self-centered, I think one says: what do they have? They have a tremendous desire to contribute. Maybe they want to succeed too fast. But how do we use what they have to make them want to belong? What is it that the non-profit institution can do to that newcomer, that young person, to acquire self-discipline?

MAX DE PREE: That's a very difficult question. I think it's better to err on the side of being more demanding of a person than of being less demanding.

PETER DRUCKER: And be willing to have a high casualty rate?

MAX DE PREE: Yes. But organizationally speaking, the casualty isn't always necessarily terminal. One of the things that I feel we need to understand better in organization life is the role of grace. Mistakes are not terminal. Mistakes are part of education with, of course, some exceptions. When we challenge people on the high side, the odds are much better that we're going to get *both* better performance and more development of the person.

PETER DRUCKER: On two conditions, I would say, Max. One has to be willing to give the person who tries a second and perhaps even third chance, but I wouldn't waste my breath on people who don't try. And then there has to be a mentor if you give that much load, that much demand, that much responsibility to beginners—and I'm all for it. I would never have learned anything if I hadn't been loaded to the gunnels by my first two bosses; they were totally unpermissive and demanding. And they did not hesitate to chastise me. But they were willing to listen to me. They were sparing with praise, but always willing to encourage. I couldn't even guess how much I owe them. I think one needs an enormous amount of responsibility, especially as a beginner, but one also does need a mentor. How do you provide that?

MAX DE PREE: In my experience, it's never been easy formally to establish mentorship programs. I think that mentorship, in a certain sense, depends on chemistry. People make a connection. One person feels ready to help another. One person feels ready to accept help from a certain person. I believe that the best way to have mentorship take place is to reward it visibly when it happens rather than to try to structure it.

PETER DRUCKER: Look out for those people—and they're not usually very conspicuous—who do a job developing people, and recognize them, praise them, feature them.

MAX DE PREE: That's right.

PETER DRUCKER: Consider it one of the key functions in the organization?

MAX DE PREE: Yes. And the leader better make sure that those people know how the leader personally feels about their contribution to the organization. That cannot slip by unnoticed.

PETER DRUCKER: Max, you have been talking about "the" leader, and yet you have been famous in your own organization for building a strong team of colleagues and conspicuous in the organizations where you were on the board for stressing the team again and again. So, what are the ways of building a team? Especially in organizations in which you have professionals on the staff and volunteers and an elected board, and so on, held together by a common mission and common vision.

MAX DE PREE: I think the first element is to understand the task. What is the job that has to be done?

PETER DRUCKER: The key activities?

MAX DE PREE: The key activities of the team. The second one is selecting people, and that's a high-risk process. When we select people, I think we have to understand that we're going to make some adjustments in assignments. Then we assign the work very clearly with a lot of interaction. We agree on what the process is going to be for getting that work done. We agree on timetables where those are appropriate. We agree on how we're going to measure performance. That all sounds fairly conventional, but it's hard work.

There's one further element: the way in which you judge the quality of leadership by what I would call the tone of the body, not by the charisma of the leader, not by how much publicity the company gets, or the leader gets, or any of that stuff. How well does the body adjust to change? How well does the body deal with conflict? How well does the body meet the needs of the constituency or customers, whatever it is? That, in the end, is the way you judge the quality of leadership.

PETER DRUCKER: Would you include in your tone of the body also what happens when that leader passes off the scene?

MAX DE PREE: Succession is one of the key responsibilities of leadership.

PETER DRUCKER: Let me try to wrap up this interview and pull it together.

We are all used to talking about the leader as the servant of the organization. And you, Max, stressed that, but you stressed something we are not hearing very often, when you talked about the indebtedness of the leader: that the leader starts out with the realization that he and the organization owe; they owe the customers, the clients, the constituency, whether they are parishioners, or patients, or students. They owe the followers, whether that's faculty, or employees, or volunteers. And what they owe is really to enable people to realize their potential, to realize their purpose in serving the organization.

5

Summary:
The Action Implications

We hear a great deal these days about leadership, and it's high time we did. But, actually, mission comes first. Non-profit institutions exist for the sake of their mission. They exist to make a difference in society and in the life of the individual. They exist for the sake of their mission, and this must never be forgotten. The first task of the leader is to make sure that everybody sees the mission, hears it, lives it. If you lose sight of your mission, you begin to stumble and it shows very, very fast. And yet, mission needs to be thought through, needs to be changed.

The basic rationale for the organization may be there for a very long time. As long as the human race is around, we'll be miserable sinners. And as long as the human race is around, we will have sick people who need to be taken care of. We know that no matter how well a society does, there will be alcoholics, there will be people in trouble with drugs, there will be people who need the Salvation Army to bring compassion to them and a little help, and an attempt to rehabilitate them, and children will have to learn and go to school. Boys and girls, as they grow up, will need scouting and experiences that form their character, that give them a role model, that give them direction and employ them intelligently so that they learn something.

We will have to look at the mission again and again to think through whether it needs to be refocused because demographics change, because we should abandon something that produces no results and eats up resources, because we have accomplished an objective. A good example is the school that is largely in crisis because it has achieved its original objective of getting every kind of child to go to school and stay there for years, and now we have to think through what we really do expect of the school. And this will be, in many ways, quite different from what the schoolmasters through the ages were striving for when nine out of ten kids never had the opportunity of organized schooling. Therefore, it is vitally important to start out from the outside. The organization that starts out from the inside and then tries to find places to put its resources is going to fritter itself away. Above all, it's going to focus on yesterday. One looks to the outside for opportunity, for a need.

At the same time, the mission is always long-range. It needs short-range efforts and very often short-range results. And yet it starts out with a long-range objective. There is a wonderful sentence in one of the sermons of that great poet and religious philosopher of the seventeenth century, John Donne: "Never start with tomorrow to reach eternity. Eternity is not being reached by small steps." So we start always with the long range, and then we feed back and say, What do we do *today?*

"Do" is the critical word. And that's the difference between what so often passes for planning in American business and what the Japanese do. It's not that they are better planners. It is that they start out by saying, Where should we be ten years hence? And we start by saying, What should be the bottom line for the quarter—which contrary to what most people in the United States believe, is higher in Japan than it is in American business, precisely because they start with the long range and feed back. As did all the companies in this country that have succeeded in staying viable, producing results for long the term. We have had some amazingly successful long-term companies—the Bell Telephone System, for fifty or sixty years; Sears, Roebuck for sixty years; General Motors, until recently. They all started out with a very clear long-range concept. Sears said: Our business is to be the informed and responsible buyer for the American family. And then one feeds back, and that may lead to very short-term moves—to Sears going into diamonds, for instance, right after World War II when the GIs came back and got married. But one always starts out with the long term. This is particularly important for non-profit institutions, precisely because they do not have an immediate bottom line, but also because they are there to serve.

But action is always short term. So one always has to ask: Is this action step leading us toward our basic long-range goal, or is it going to sidetrack us, going to divert us, going to make us lose sight of what we are here to do? This is the first question.

But also we need to be result-driven. We need to ask, Do we get adequate results for our efforts? Is this their best allocation? Yes, need is always a reason, but by itself it is not enough. There also have to be results. There also has to be something to point at and say, We have not worked in vain. So we are always looking at programs and projects with the question, Do they produce the right results? The leader's job is to make sure the right results are being achieved, the right things are being done.

One has the responsibility to allocate resources, particularly of course in organizations that depend heavily on volunteers, and heavily on donors. Leadership is accountable for results. And leadership always asks, Are we really faithful stewards of the talents entrusted to us? The talents, the gifts of people—the talents, the gifts of money. Leadership is *doing*. It isn't just thinking great thoughts; it isn't just charisma; it isn't play-acting. It is doing. And the first imperative of doing is to revise the mission, to refocus it, and to build and organize, and then abandon. It is asking ourselves whether, knowing what we now know, we would go into this again. Would we stress it? Would we pour more resources in, or would we taper off? That is the first action command for any mission.

It is also the one way of keeping an organization lean and hungry and capable of doing new things. An old medical proverb says that the body can only take in the new if it eliminates the waste products. This is therefore the first action requirement: the constant resharpening, the constant refocusing, never really being satisfied. And the time to do this is when you are successful. If you wait until things have already started to go down, then it's very difficult. It is not impossible to turn around a declining institution, but an ounce of prevention is very much better than a ton of cure in the turnaround situation.

The next thing to do is to think through priorities. That's easy to say. But to act on it is hard because it always involves abandoning things that look very attractive, that people both inside and outside the organization are pushing for. But if you don't concentrate your institution's resources, you are not going to get results. This may be the ultimate test of leadership: the ability to think through the priority decision and to make it stick.

Leadership is also example. The leader is visible; he stands for the organization. He may be totally anonymous the moment he leaves that office and steps into his car to drive home. But inside the organization, he or she is very visible, and this isn't just true of the small and local one, it is just as true of the big, national, or worldwide one. Leaders set examples. The leaders have to live up to the expectations regarding their behavior. No matter that the rest of the organization doesn't do it; the leader represents not only what we are, but, above all, what we know we should be.

So it is a very good rule when you do anything as a leader, to ask yourself, Is that what I want to see tomorrow morning when I look into the mirror? Is that the kind of person I want to see as my leader? And if you follow that rule, you will avoid the mistakes that again and again destroy leaders: sexual looseness in an organization that preaches sexual rectitude, petty cheating, all the stupid things we do. Maybe the individual does them; well, that's his or her business. But a leader is not a private person; a leader represents. And then ask yourself, as a leader, what do I do to set standards in the organization? What do I do to enable the organization to tackle new challenges, to seize new opportunities, to innovate? What do *I* do? Not what does the organization do? Take action responsibility. What are my own first priorities, and what are the organization's first priorities, what *should* they be? These are the action agenda. These are the things that must be done.

You may think, that's fine for the CEO, but I'm only a volunteer putting in three hours a week, being a den mother or arranging flowers at a patient's bedside. *You are a leader.* The exciting thing, the new thing, is that we are creating a society of citizens in the old sense of people who actively work, rather than just passively vote and pay taxes. We are not doing it in business. There is a lot of talk of participative management; but there is not much reality to it, and in many ways, there never will be. The pressures are perhaps too great. In a country like ours, with almost 250 million people, even a small town has 50,000 inhabitants, and there is not very much a citizen can actually *do.* We could not, even in the smallest town, meaningfully revive the New England town meeting of two hundred years ago, when that New England town had one hundred twenty people or so.

But we are doing exactly this in the non-profit, the service institution, where increasingly there are only leaders. These are people who are paid and people who are not paid. In a church there are a very small number of people who are ordained, but one thousand people who work and do major tasks for the church who are not ordained, never will be, never get a penny. In the Girl Scouts of the U.S.A., there are one hundred volunteers for every paid staff member, and each is doing a responsible task. We are creating tomorrow's society of citizens through the non-profit service institution. And in that society, everybody is a leader, everybody is responsible, everybody acts. Everybody focuses himself or herself. Everybody raises the vision, the competence, and the performance of his or her organization. Therefore, mission and leadership are not just things to read about, to listen to. They are things to do something about. Things that you can, and should, convert from good intentions and from knowledge into effective action, not next year, but tomorrow morning.

PART TWO

From Mission
to Performance

*effective strategies for marketing,
innovation, and fund development*

1

Converting Good Intentions into Results

The non-profit institution is not merely delivering a service. It wants the end user to be not a user but a *doer*. It uses a service to bring about change in a human being. In that sense a school, for instance, is quite different from Procter & Gamble. It creates habits, vision, commitment, knowledge. It attempts to become a part of the recipient rather than merely a supplier. Until this has happened, the non-profit institution has had no results; it has only had good intentions.

Napoleon said that there were three things needed to fight a war. The first is money. The second is money. And the third is money. That may be true for war, but it's not true for the non-profit organization. There you need four things. You need a plan. You need marketing. You need people. And you need money.

The plan we have just talked about, in the first part. People we will be talking about a little later, in this book's fourth and fifth parts. In this part we talk about the *strategies* that convert the plan into results. How do we get our service to the "customer," that is, to the community we exist to serve? How do we market it? And how do we get the money we need to provide the service?

Non-profit institutions that do well used to think they didn't need marketing. But, as a famous old saying by a great nineteenth-century con man has it, "It's much easier to sell the Brooklyn Bridge than to give it away." Nobody trusts you if you offer something for free. You need to market even the most beneficial

53

service. But the marketing you do in the non-profit sector is quite different from selling. It's more a matter of knowing your market—call it market research—of segmenting your market, of looking at your service from the recipient's point of view. You have to know what to sell, to whom to sell, and when to sell. Although marketing for a non-profit uses many of the same terms and even many of the same tools as a business, it is really quite different because the non-profit is selling something intangible. Something that you transform into a value for the customer. The sick patient in the hospital doesn't have to be sold. You are not marketing the sickness of that patient to the physician, who is the non-profit hospital's main customer. You are marketing what you can do to help the physician in his or her practice. That's a concept—an abstraction—and to sell a concept is different from selling a product.

To run a non-profit effectively, the marketing must be built into the design of the service. This is very much a top management job, although, as in every other area, you need a lot of input from your people, from the market, and from research. A big national organization, such as the American Cancer Society, for instance, probably has the most elaborate market research, using detailed census data for fund-raising, a physicians' advisory committee to work directly with physicians who are in many ways its first market, and so on. The American Cancer Society doesn't design a service and then start peddling it.

That uniquely American invention, the Community Chest—or the United Way, as it is often called—is in many ways a response to the market. People got awfully tired of being hit for a donation by twenty-nine different organizations and became suspicious that this meant exceedingly high collection costs, with most of the money going into doorbell ringing rather than feeding the hungry. The design of the United Way hasn't changed much over the years: the employers of the community are its collection agents. But the United Way has to keep its marketing up to date, has to adjust to the changing business population, has to know which employers to go to and which local societies to bring onto its board, so that it can work effectively with industry. It has to understand the changing structure of employment to design its most effective appeal. The non-profits who don't do that, who think they can rely on high-pressure selling, just don't do very well.

An important point to remember, incidentally, in designing a non-profit's service and marketing is to focus only on those things you are competent to do. If you run a hospital, you'd better not try to do what you are not competent to do. For clinical neurology, you need a certain critical mass—forty beds, fifty beds—to do a decent job. If you are the only hospital in Silver Fish, South Dakota, and there's not another hospital around for one hundred miles, you have to do what has to be done. Even there, let me say, you probably would do better to fly that neurology patient by helicopter to the nearest medical center—not for financial but for competence reasons. You know the general advice is, don't go to a hospital to have a heart by-pass if they don't do two or three hundred by-passes a year. You have to do these very demanding technical things again and again—and again. The same is true of colleges. In fact, a great disease of the liberal arts college is that it thinks it can do everything. Don't put your scarce resources where you aren't going to have results. This may be the first rule for effective marketing.

And then, the second rule, know your customers. Yes, I said *customers*. Practically everybody has more than one customer, if you define a customer as a person who can say no. When you look at the soap manufacturer, the supermarket doesn't have to put a manufacturer's detergent on the shelf, and certainly not in a position where the housewife will see it. And yet, unless the housewife also wants to buy the detergent, you have no sale, so you have two customers. The Boy Scouts or Girl Scouts have even more customers: they have the parents, and they have the kids. But then there are the volunteers, without whom no scouting organization could be run. And the teachers in the school also have to be "sold" on scouting, or they could easily impede or perhaps even veto it.

So, the design of the right marketing strategy for the non-profit institution's service is the first basic strategy task: the non-profit

institution needs market knowledge. It needs a marketing plan
with specific objectives and goals. And it needs what I call marketing responsibility, which is to take one's customers seriously. Not
saying, We know what's good for them. But, What are their values? How do I reach them?

The non-profit institution also needs a fund development strategy.
The source of its money is probably the greatest single difference
between the non-profit sector and business and government. A
business raises money by selling to its customers; the government
taxes. The non-profit institution has to raise money from donors.
It raises its money—at least, a large portion of it—from people
who want to participate in the cause but who are not beneficiaries.

   Almost by definition, money is always scarce in a non-profit
institution. Indeed, a good many non-profit executives seem to
believe that all their problems would be solved if only they had
more money. In fact, some of them come close to believing that
money-raising is really their mission. An example is some presidents of private colleges or universities who are so totally preoc-
cupied with money-raising that they have neither the time nor the
thought for educational leadership.

   But a non-profit institution that becomes a prisoner of money-
raising is in serious trouble and in a serious identity crisis. The
purpose of a strategy for raising money is precisely to enable the
non-profit institution to carry out its mission without subordinat-
ing that mission to fund-raising. This is why non-profit people
have now changed the term they use from "fund raising" to "fund
development." Fund-raising is going around with a begging bowl,
asking for money because the *need* is so great. Fund development
is creating a constituency which supports the organization because
it *deserves* it. It means developing what I call a membership that
participates through giving.

   Your first constituency in fund development is your own board.
One of the things we have learned about managing non-profit
institutions is that the old-type board, the board that simply was
in sympathy with the institution, is no longer enough. You need

a board that takes an active lead in raising money, whose members give both of themselves and by being fund-raisers, fund *developers*. When a board member calls, say, a real estate developer, and says, "I am on the board of the hospital," the first response he gets from his friend is, "How much are you giving yourself, John?" If the answer is five hundred bucks, well, that's all you're likely to get.

But you also want something else on the board which has to do with money: the ability to audit the balance between your program and your resources. That is what gives you assurance. The person who runs the church or hospital or school should be enthusiastic. You don't want nay-sayers in those positions. But somebody has to ask: "Is this the best balance between our available resources and our effectiveness?"

A business earns its money on its own. The money of the non-profit institution is not its own; it is held in trust for the donors. And the board is the guardian to make sure the money is used for the results for which it has been given. That, too, is part of the non-profit strategy.

Not so long ago, many non-profit organizations were pretty self-supporting financially. They generally needed outside money only for extra projects—that new science hall, or a new cardiac wing. Now, more and more non-profit organizations need money for operating purposes. Another reason why the development of financial resources is becoming more important is because great wealth is becoming less important. It used to be that two or three rich people in the community supported the Church. That doesn't work anymore. Not only is the Church more expensive, but demands on people of great wealth have gone up out of sight. And, proportionately, there are so many fewer of them around. So, non-profit executives must build a mass base.

You need people on your board willing to help develop that mass base by giving example and leadership.

Of course, there will always be need for emergency relief and appeals to give for it—for the most recent earthquake, for starving children in Africa or the Vietnamese boat people. But it is increasingly dangerous to depend on emotional appeal alone. A friend of mine who heads a major international relief organization speaks of "compassion fatigue." There is so much misery in the world that we are becoming quite hardened and callous to that constant plucking of our heart strings.

In fund development you appeal to the heart, but you also have to appeal to the head, and try to build a continuing effort. The non-profit manager has to think through how to define *results* for an effort, and then report back to the donors, to show them that they are achieving results.

You also have to educate donors so that they can recognize and accept what the results are. This is perhaps the newest development—this realization that a donor doesn't automatically understand what the organization is trying to do. Donors are becoming too sophisticated to appeal to them simply on the basis that education is good or health is good. They ask, Whom are you educating? Educating for what?

This moves us to constituency building over the very long term. It is how the Claremont Colleges, where I have been teaching now for twenty years, were built. In the 1920s, the president of Pomona College, the mother college of the group, realized that Southern California and its college population would grow fast and that he would need a great deal of money for the college. He started by actually founding local new businesses and running them for a couple of years until they broke even. Then he called in a top-flight new graduate, literally gave him the business and $10,000 to boot (which was a great deal of money in those days), and said, "It's yours. You build it. But if it is successful, don't repay us. Remember us." That's why Pomona and the whole Claremont group are so well endowed today. He built an enormous constituency—long term. The fruits didn't come in for twenty years, but they came in a thousandfold. I'm not saying that this is the way everyone should do it. But it is one example of building up a long-term constituency, people who remember, who are not giving simply because someone rings a doorbell. They see the support of the institution as self-fulfillment. That is the ultimate goal of fund development.

2

Winning Strategies

There is an old saying that good intentions don't move mountains; bulldozers do. In non-profit management, the mission and the plan—if that's all there is—are the good intentions. Strategies are the bulldozers. They convert what you want to do into accomplishment. They are particularly important in non-profit organizations. One prays for miracles but works for results, St. Augustine said. Well, strategies lead you to work for results. They convert intentions into action and busyness into work. They also tell you what you need to have by way of resources and people to get the results.

I was once opposed to the term "strategy." I thought that it smacked too much of the military. But I have slowly become a convert. That's because in many businesses and non-profit organizations, planning is an intellectual exercise. You put it in a nicely bound volume on your shelf and leave it there. Everybody feels virtuous: We have done the planning. But until it becomes actual work, you have done nothing. Strategies, on the other hand, are action-focused. So, I've reluctantly accepted the word because it's so clear that strategies are not something you *hope* for; strategies are something you *work* for.

Here is one example of a winning strategy: Brown University in Providence, Rhode Island. Twenty years ago it was a respectable "also ran," known as "Harvard's little sister." It had an excellent faculty. But it had no distinction; it did what everybody else did. Then a new president asked, What do we have to do to become a leader despite the tough competition where we have Harvard to the north and Yale to the south, and about twelve first-rate liberal

arts colleges within an hour's drive? He focused on two things.
First, make women full citizens of the university. Brown always
had a women's college—Pembroke. But making women full citizens meant bringing in those women who wanted to go where
women supposedly don't go—mathematics, the sciences, pre-med,
computers—and systematically recruiting young women who
were doing exceptionally well in these areas that tradition doesn't
consider particularly feminine. Second, build closeness to students
into the way the university runs. For each of these two goals the
new president had a strategy. In the past ten years, Brown has
become the "in" university for bright kids in the East.

This is almost a textbook case of a successful *marketing* strategy. What that Brown president did was to recognize changes in
the market: the emergence of career-focused young women, and
the desire of students, after the turbulent sixties, to have a "community." And then he developed specific campaigns to reach his
potential customers—and went to work.

IMPROVING WHAT WE ALREADY DO WELL

In this country, in particular, we usually underplay the strategy
of doing better what we already do well. This hit me the first time
I went to Japan, when they were just beginning their meteoric rise.
I looked for innovation strategies and there weren't any. But every
place—whether university, business, or government agency—had
a clear strategy for *improving*. They don't talk innovation. They
ask, How do we do better what we are already doing? It may be
something very mundane, like sweeping the floor. Or it may be a
very major change: don't just bring in new machines and put them
into the existing lineup; we really have to change the lineup and
rearrange the whole process. But the focus is always on improving
the product, improving the process, improving the way we work,
the way we train. And you need a continuing strategy for doing
so.

To work systematically on the productivity of an institution, one needs a strategy for each of the factors of production. The first factor is always people. It's not a matter of working harder; we learned that long ago. It's a matter of working smarter, and above all, of placing people where they can really produce. The second universal factor is money. How do we get a little more out of the money that we have? It's always scarce. And the third factor is time.

One needs productivity goals—and ambitious ones. Whenever I sit down with people to discuss productivity goals, they say, "You are way too high." I learned from an old friend, one of the great men of the black community, Kenneth Clark, the psychologist at City University of New York, that one should always set the objective twice as high as one hopes to accomplish because one will always fall by 50 percent short. That's a little cynical, but there's truth in it, so set your objectives high. Not so high that people say this is absolutely absurd, but high enough so that they say: we've got to stretch.

Constant improvement also includes abandoning the things that no longer work; and it includes the innovation objective. Let's take 3M, which turns out two hundred new products a year. They start out by saying that 80 percent of the products that will be in the line ten years hence we haven't even heard of yet. And then they go to work, to work, to work. Almost everything human beings use becomes obsolete sooner or later, so we have to replace it. What is our innovation strategy? Where are we going to do something different, or do the same thing quite differently? Set the goals—and go to work.

For non-profit managers, the signposts are less clear. How, for instance, in a mental-health clinic can you judge the effectiveness of a strategy, whether you're doing better this year than last?

Well, you can define what "better" means. I know one major mental-health clinic that does a tremendous job in an area in which results are terribly hard to achieve—mostly paranoia cases. The head of that clinic is a good friend, and I said to him, "Working with paranoia must be terribly frustrating. In depression, we can help people today. In schizophrenia, we can help, not perhaps a great many, but quite a few. But with paranoia, there are very limited results." He answered, "You are wrong, Peter. We have a simple goal. We know we don't know how to cure it; we don't understand it at all. But there is a possibility of helping people who are sick with paranoia to realize that they are sick. And that is a tremendous step forward. Because then they know that *they* are sick, not that the world is sick. They are not cured, but they function."

That's a qualitative goal. You can set goals that are not measurable but can be appraised and can be judged.

The people in a really successful research laboratory cannot quantify their research results ahead of time. But they can sit down every three years and ask, What have we contributed in the last three years that did make a difference? And what do we plan to contribute? These are qualitative measures. And they are just as important as the quantitative ones. Let me say that you have to define quality first. Quantity without quality is the worst thing and will result in total failure.

And how does a pastor set a strategy? First, the pastor has to define the goals. What is he or she trying to do? Sure, you make certain assumptions about people if you are a pastor. You make the assumption that it probably increases their chance of salvation if they do go to church.

As an old schoolmaster, which I've been for sixty years, I assume that the longer kids sit on their backsides, the more they learn. These are not testable assumptions, but you've got to make them. So the pastor sets his or her goal, which is to build a congregation.

What kind of a congregation? Not every pastor has the same vision. You may find people who say, I just bring them to church; that's the main goal. The next one will say, No, I only want to bring certain kinds of people in. Both are in the same profession, but see their mission quite differently. One sees it as to build a broad base; the other wants to start a small community of true believers who will stand fast at Armageddon.

Then you have to ask, What are the specific results I want?

Whether it's a church or a hospital or a Boy Scout troop or a public library, your strategy will have the same structure. First, you need the goal, and it's got to fit your mission. But it also has to fit the environment in which you work. Then you think through specific results for specific areas. The pastor who sees his or her church in terms of large masses of parishioners segments the market and designs a service for each segment. I once sat in on a meeting at which a very successful pastor said: "Any fool who is ordained can build a large church in five to seven years if he focuses on five market segments. He has a youth ministry, a singles ministry, a young-married ministry, a home ministry for the shut-ins, and a ministry for the elderly. The rest is hard work." Then he added, "Of course, the targets you set for these five depend on the community you are in."

That's perhaps a little oversimplified, but I've heard hospital administrators talk very similarly. Look at the ultimate beneficiaries—call them the market—the ultimate clients. Whether that market is a church, a hospital, a Boy Scout troop, or a public library, you have the same structure for your strategy. If you are a public library, you have adults, young people, and preschoolers, and you serve the schools. I think of each of these groups as a separate market—they share a building, they share common services, they share a lot of books; but I think you go after them separately. And you develop a marketing plan. You will need money, and will have to allocate it sensibly. You will have to communicate and you will have to have feedback.

First, the goal must be clearly defined. Then that goal must be converted into specific results, specific targets, each focused on a specific audience, a specific market area. You may need a great many such specific strategies. The American Heart Association divides the American public from which it raises money into forty-one different segments. That's quite a lot. But it explains perhaps why they have been so successful.

Next, you will need a marketing plan and marketing efforts for each target group. How are you really going to reach this specific segment? You now need resources—people, above all—and money. And the allocation of both.

Next comes communication—lots of it—and training. Who has to do what, when, and with what results? What tools do they need? In what language do they have to hear it? One pastor told me that when he sits down with each of his groups and talks about goals and missions, even after twenty-five years, he still uses the language of the seminary. But the people who have to do the work are his lay volunteers, and to them these are strange words. Words like "implementation," "fulfillment," and "plan" when used by a hospital administration may sound strange, too, for somebody in physical therapy who knows all the muscles of the body. You have to ask who must do what, and in what form they should get it so that it becomes *their* work.

Then you need logistics—for want of a better word. What resources are required? I'm always reminded of the old story that whenever Napoleon's brilliant marshals came out with great plans of moving against Prussia, or Spain, or what have you, Napoleon would listen silently and then ask, "How many horses does it require?" Usually they hadn't thought it through and their plan outran the available horses. That's very typical.

Finally, you ask: "When do we have to see results?" Try not to be impatient. But you must be able to see whether you are on course when the results come in. What feedback do you need? How do you measure your achievement so that you realize that in this area, which is crucial, we are way behind our timetable? If we can't speed it up, we will have to scale everything back (the horses aren't there, in Napoleon's terms). Or, here we are ahead of our timetable? Is this an opportunity for us to speed it all up, or does it indicate that we are getting dangerously out of line? You need feedback and control points.

I think the steps are the same for every organization. How you carry them out depends very much on what kind of an organization you are.

To carry out the process, you need to use both written and verbal communication. A written process has the great advantage that you can hand out a sheet to everybody, go down the line, check it off, and say, "Any questions on point three?" And somebody says, "Are we on point three? I thought we were still on point two." You talk about it. Above all, you invite questions.

But you also have to encourage people to come back and say, "This is what I heard. Am I right that you expect me to do this?" That is much better done in speaking than in writing. Partly because there is less misunderstanding and partly because it's freer and less formal.

To my mind, the best example of a winning strategy in a nonprofit institution is that of The Nature Conservancy. Its clear goal is to preserve as much as possible of God's ecological diversity of flora and fauna, which is endangered by man. The board members developed one strategy to find the places that needed preserving; another to get the money to buy them; and a third to manage it. The market—the people from whom they get the money—is local. So they built state organizations that reach the local people, and a goal of, I believe, fifteen of these major nature preserves per year, which is very ambitious. They are meeting it because they were so clear about that goal and its implementation. I think that accounts for success pretty much across the board.

But there is one *don't* on strategy. Don't avoid defining your goals because it might be thought "controversial." This strategy almost destroyed a major hospital, which attempted to brush under the rug one tough and highly controversial question: Are we trying to fill as many of our beds as possible, or are we trying to deliver the best patient care? Their best-known eye doctors had proposed to move eye surgery into a free-standing ambulatory eye clinic, next door to the hospital. The eye surgeons saw this as improving patient care, indeed, as the first step toward the healthcare delivery system of tomorrow; and so did some of the hospital administrators. But the board saw only that this move would cut bed occupancy, which was their first priority. Eventually, the prominent eye surgeons got tired of the wrangle and left the hospital altogether, taking with them their patients, both ambulatory and bed patients. Other prominent physicians followed. Three

years later the hospital had gone downhill so much—both in reputation and in occupancy rate—that it had to sell out to a for-profit
chain.

With strategy, one always makes compromises on implementation. But one does not compromise on goals, does not pussy-foot
around them, does not try to serve two masters.

Here is another *don't:* Don't try to reach different market segments with the same message. Some years ago I helped develop an
executive management program. We were crystal-clear on the
goal; but we did not really spend enough time thinking through
the market segments. We tried to sell the program to everybody
the same way. After six or seven years of working very hard and
not getting very far, we sat down and said, "Look, we really have
three quite separate markets. They may all belong in the same
program, but they are coming for different reasons." We organized
it to the point that we now have different administrators for these
groups. And it works.

HOW TO INNOVATE

Usually, there is no lack of ideas in non-profit organizations.
What's more often lacking is the willingness and the ability to
convert those ideas into effective results. What is needed is an
innovative strategy. The successful non-profit organization is organized for the new—organized to perceive opportunities. Innovative organizations systematically look both outside and inside for
clues to innovative opportunities.

One strategy is practically infallible: Refocus and change the
organization *when you are successful.* When everything is going
beautifully. When everybody says, "Don't rock the boat. If it ain't
broke, don't fix it." At that point, let's hope, you have some
character in the organization who is willing to be unpopular by
saying, "Let's *improve* it." If you don't improve it, you go downhill
pretty fast.

The great majority of major institutions that have gotten into real trouble over the last fifteen years are successes that rested on their laurels. Look at the American labor union in the early Eisenhower years. It was king of the jungle. But where is it now? The main reason for these calamities is that the people who said then, "We have obtained our objectives; now let's improve on them," were forced out. They were treated like the little boy who says a dirty word in church. Sears, Roebuck twenty years ago was the merchant prince, the first choice of seventy percent of American families. It became so complacent that it ignored all signs of change in the American marketplace. When you are successful is the very time to ask, "Can't we do better?" The best rule for improvement strategies is to put your efforts into your successes. Improve the areas of success, and change them.

The responsibility for this rests at the top, as in everything that has to do with the *spirit* of an organization. And so the executives who run innovative organizations must train themselves to look out the window, to look for change. The funny thing is, it's easier to learn to look *out* the window than to look *inside,* and that's also a smart thing to do systematically.

The most successful college I know has managed—at a time of shrinking student population—to increase the number of its applicants and improve the quality of those applicants by just such a discipline. The president and the director of admissions spend alternate weeks visiting high schools and inquiring about the changing expectations of the kids. The pastoral church which is such a significant sociological phenomenon in today's America looks at changes in demographics, at all the young, professional, educated people who have been divorced from their background and need a community, need help, comfort, and spiritual sustenance. The change outside is an opportunity. You can force yourself to drive a different route to work; you can force yourself to sit down and talk with students, who are still in high school but thinking of college. You can force yourself to look at demographics—and that's your first source.

Then you look *inside* your organization and search for the most important clue pointing the way to change: generally, it will be the unexpected success. Most organizations feel that they deserve the unexpected success and congratulate themselves on it. Very few see it as a call to action. My best story on this score is not an American story, it's from India, which has converted itself in less than twenty years from chronic famine to food surplus. One of the keys to this change was the unusual success of a large farm cooperative that had become the sales agent for a cheap European bicycle with an auxiliary motor. The only trouble was the farmers didn't want it; they didn't buy it. Amazingly enough, while orders for bicycles didn't come in, orders for *replacement* motors for bicycles the farmers hadn't bought came in by the bushel. Everybody said, "Those stupid farmers, don't they know they need a bicycle?" Except for one co-op official. He went out and asked, "What are you doing with them?" Well, he found that the farmers were using that little single-stroke gasoline engine as a motor for irrigation pumps, which had always been powered by hand. Perhaps the greatest single contribution to India's tremendous agricultural success is the gasoline irrigation pumps that now bring available water to where it's needed.

The first requirement for successful innovation is to look at a change as a potential opportunity instead of a threat.

Everybody is worried about the "latchkey kids." But for the Girl Scouts of the U.S.A., the fact that there are so many young girls today whose mothers are out at work became a tremendous opportunity that led to the creation of the Daisy Scouts. Faced with a change, we should always ask, How can this give us a chance to contribute?

The second question is, Who in our organization should really work on this? That's a crucial question. Most new things need to be incubated. They need to be piloted by somebody who really wants that innovation, who wants it to grow, who believes in it. Everything new also gets into trouble, so look for somebody who really wants to commit himself or herself and who has enough standing in the organization.

Then think through the proper marketing strategy. What are you really trying to do? When you look at successful businesses,

they have very different strategies. A company like Procter &
Gamble has always had one clear strategy in bringing out a product: to be the first, and to dominate the market. If it works, that's
a winning strategy; but it's terribly risky. For fifty years, IBM
never has brought out a new product; it has always been a creative
imitator. It also always aimed at market dominance, but it lets
somebody else go in first because the first version is likely not to
be quite right. The Japanese strategy is very different. It exploits
the mistakes of the leaders, their bad habits, especially their arrogance.

Look into the possibility of developing a niche. One very successful non-profit hospital group does not develop community
hospitals but examines what each local community needs. In one
community there is room for a psychiatric hospital, in another
room for a good gerontology center. Each is a specialty hospital.
That is a strategy: if you come out with a specialty, don't try to
do everything for everybody.

THE COMMON MISTAKES

There are a few common mistakes in doing anything new.

One is to go from idea into full-scale operation. Don't omit
testing the idea. Don't omit the pilot stage. If you do, and skip
from concept to the full scale, even tiny and easily correctible flaws
will destroy the innovation.

But also don't go by what "everybody knows" instead of looking
out the window. What everybody knows is usually twenty years
out of date. In political campaigns, the ones who look so promising
at the beginning and then fizzle out are usually the ones who go
by what they believe everybody knows. They haven't tested it, and
it turns out that "This was twenty years ago."

The next most common mistake is righteous arrogance. Innovators are so proud of their innovation that they are not willing to
adapt it to reality. It's an old rule that everything that's new has
a different market from the one the innovator actually expected.

I remember one of my pastoral friends saying of a new program: "Great, a wonderful program for the newly married." The program was indeed a success. But to the consternation of the young assistant pastor who had designed it and ran it, not a single newly married couple enrolled in it. All the participants were young people who were living together and wondering whether they should get married. And as the senior pastor told me, he had a terrible time with his brilliant young assistant, who became righteous and said, "We haven't designed it for them." He wanted to throw them out.

Another common mistake is to patch up the old rather than to go all-out for the new. The present plight of General Motors clearly shows that in that case you will get only the costs of the new, and none of its benefits. When the Japanese came in and the American public changed its approach to the automobile, GM patched. GM improved a little bit what it already was doing, and spent enormous amounts of money and time and people on patching—far more than genuine innovation would have required. A few years later Ford ran around GM. Ford sat down and said, "What does the new require?" It designed new cars and new ways of selling them, and risked a good deal of existing investment. But Ford brought out something that looked and behaved differently, and that could really compete.

There comes a point when one has to look at what the job requires, and design for that job, rather than saying, "This is how we've always done it. Let's improve it a little bit." This is one of the critical decisions. It is one of the crucial tasks of the executive to know when to say, "Enough is enough. Let's stop improving. There are too many patches on those pants."

Don't assume that there is just the one right strategy for innovations. Every one requires thinking through anew. Don't say, "We have been successful six times in introducing the new this way, so that must be the *right* way. That's our formula now." And, if it doesn't work, don't blame the "stupid public." Say instead, "Maybe this needs to be done differently." Before you go into an innovative strategy, don't say, "This is how we do it." Say, "Let's find out what this needs. Where is the right place in the market? Who are the customers, the beneficiaries? What is the right way to deliver it? What is the right way to introduce it? Let's not start out with what we know. Let's start out with *what we need to learn.*"

When a strategy or an action doesn't seem to be working, the rule is, "If at first you don't succeed, try once more. Then do something else." The first time around, a new strategy very often doesn't work. Then one must sit down and ask what has been learned. "Maybe we pushed too hard when we had success. Or we thought we had won and slackened our efforts." Or maybe the service isn't quite right. Try to improve it, to change it and make another major effort. Maybe, though I am reluctant to encourage that, you should make a third effort. After that, go to work where the results are. There is only so much time and so many resources, and there is so much work to be done.

There are exceptions. You can see some great achievements where people labored in the wilderness for twenty-five years. But they are very rare. Most of the people who persist in the wilderness leave nothing behind but bleached bones. There are also true believers who are dedicated to a cause where success, failure, and results are irrelevant, and we need such people. They are our conscience. But very few of them achieve. Maybe their rewards are in Heaven. But that's not sure, either. "There is no joy in Heaven over empty churches," St. Augustine wrote sixteen hundred years ago to one of his monks who busily built churches all over the desert. So, if you have no results, try a second time. Then look at it carefully and move on to something else.

3

Defining the Market

*Interview with Philip Kotler**

PETER DRUCKER: Philip, when you published your book *Non-profit Marketing* in 1971 — it's now called *Strategic Marketing for Non-Profit Institutions* and is in its fourth edition—there was no awareness at all, am I right, on the part of non-profit institutions that they have to market and little receptivity for the idea?

PHILIP KOTLER: That's true. They were interested in doing a better job of accounting and finance, and your ideas in management were beginning to be used by them. But they had not talked about marketing. In fact, my observation is that some of them *were* doing it but without any awareness of how to do it well. I felt very strongly that marketing, like the other business functions, was generic and universal, and applied to all institutions, and that it ought to be brought into the non-profit world more consciously.

PETER DRUCKER: Since then, a good many non-profit institutions have accepted the need in theory. By and large are they translating it into practice?

PHILIP KOTLER: Different institutions took to marketing at different rates. Hospitals certainly recognized the importance of the

---

*Philip Kotler teaches at the J. L. Kellog Graduate School of Management of Northwestern University in Evanston, Illinois. His pioneering work, *Strategic Marketing for Non-Profit Institutions,* first published in 1971, is now in its fourth edition.

marketing functions, but colleges are somewhat behind. Museums and the performing arts have taken to marketing. Many institutions misunderstand it. They confuse marketing with either hard selling or advertising, and therefore, don't show an aptitude for it.

PETER DRUCKER: Well, then, how would you define marketing, especially in the non-profit institution? Most of my friends in non-profit, I think, would be somewhat nonplussed by what you just said, that they confuse it with hard selling or advertising. Most of them think that's precisely what marketing is.

PHILIP KOTLER: The most important tasks in marketing have to do with studying the market, segmenting it, targeting the groups you want to service, positioning yourself in the market, and creating a service that meets needs out there. Advertising and selling are afterthoughts. I don't want to minimize their importance. But you put it so well years ago when you shocked a number of people by saying that the aim of marketing is to make selling unnecessary.

What could marketing be if it isn't selling? The shortest definition I've heard is that it is finding needs and filling them. I would add that it produces positive value for both parties. The contrast between marketing and selling is whether you start with customers, or consumers, or groups you want to serve well—that's marketing. If you start with a set of products you have, and want to push them out into any market you can find, that's selling.

PETER DRUCKER: A good many of my non-profit friends would agree with what you just said wholeheartedly. Then they would say, But isn't the need we serve obvious? There are people who are poor and somebody has to fill their stomachs. There are people out there who live in sin and somebody has to bring the Spirit to them. They consider themselves need-driven and they don't quite understand why you have to do anything else. Is that a very one-sided view?

PHILIP KOTLER: Many organizations are very clear about the needs they would like to serve, but they often don't understand these needs from the perspective of the customers. They make assumptions based on their own interpretation of the needs out there. Let's take a hospital. The question often arises, Is it a sickness institution or a wellness institution? Most hospitals say that they are there to take care of people who are sick and to make them well. You could also argue that their real mission would be more meaningful if they set up to prevent illness. There are a lot of subtleties about needs that require interpretation and what I call customer research, consumer research. Basically, the issue is, are these organizations consumer-minded?

PETER DRUCKER: Can you give me an example of a non-profit institution that understands marketing and practices it? What do they do?

PHILIP KOTLER: Stanford University and the way it raises money from its alumni and others. Stanford took a totally market-oriented point of view. Their development office was organized with managers at the head of different alumni groups. Each group is reached in the most cost-effective way. For example, those who graduated from Stanford receive two letters, direct-mail pieces, before the university gives up on them. Those who give $25 to $75 receive three to four more letters. Those who give a little more than $75 get a telephone call, and so on. Basically, the whole development effort is based on segmenting the market and working up the most cost-effective marketing mix of tools for raising money.

PETER DRUCKER: Did Stanford do any customer research to see what the potential donors value in a university? Or did it simply say, as most colleges say, Education is good, we need your money?

PHILIP KOTLER: True, that is the problem with many sales-oriented or product-oriented organizations that think they have such a good product, they don't understand why people are not rushing to buy it or to use it. In Stanford's case, they have approached their fund-raising experimentally. They don't feel there's a universal appeal that you can make to all Stanford graduates. Different strokes for different folks works a lot better. They learned the best strokes by getting feedback and researching each market.

PETER DRUCKER: Stanford has to recruit students. That's one marketing effort. It has to attract and hold first-rate faculty, that is, people who could go to twenty other schools. And it has to develop donors and raise money. That's equally a marketing effort. You don't see any difference between the three, basically?

PHILIP KOTLER: Every organization is swimming in a sea of publics. A college will want to attract students. It will want to attract research money from government and other sources. The problem marketing has to solve is, How do I get the response I want? The answer marketing gives is that you must formulate an offer to put out to the group from which you want a response. The process of getting that answer, I call exchange thinking. What must I give in order to get? How can I add value to the other party in such a way that I add value to what I want? Reciprocity and exchange underlie marketing thinking.

PETER DRUCKER: And how important is it in this approach for the non-profit institution to differentiate itself? Stanford probably has a couple of hundred other colleges in competition; a local hospital may have three other hospitals in the same area. How important is it to differentiate yourself? And how do you do it?

PHILIP KOTLER: Marketing is now thought of as a process of segmenting, targeting, and positioning—I call it STP marketing. That's opposed to LGD marketing—lunch, golf, and dinner marketing, which may have its place, but it's not the same as doing the right job of segmenting, targeting, and positioning.

Positioning raises the question: How do we put ourselves across to a market we're interested in? How do we stand out in some way? You cannot be all things to all people. So most organizations engage in the search for their own uniqueness, what we might call a competitive advantage or advantages. That comes by cultivating certain strengths and putting them across as meaningful to the market you're going after. Let me give you an illustration. A hospital could offer the normal range of services to the patients, but, in that regard, may not be different from any other hospital. What I've seen hospitals do is identify needs in the community that were not being satisfied. For example, there may be no sports medicine program; there may be no burn unit; and so on. If the managers of these hospitals are smart, they figure out which of these needs are strong needs or which ones they can serve well. By addressing those needs, the hospital actually adds a crown jewel to itself. It adds a point of distinction. Differentiation must proceed that way. Otherwise, the customer has no reason for the choices that are going to be made.

PETER DRUCKER: So, one of the first steps in marketing for the non-profit institution is to define its markets, its publics. Think through to whom you have to market your product and your strengths. That really comes before you think through the message, does it?

PHILIP KOTLER: Yes. Let's take churches, for example, because what you said poses a real problem for churches. On the one hand, a church should go after every person who wants religious experience, and so on. It should therefore be a very diverse institution. On the other hand, marketing would suggest that it would be more successful if it defined its target group, whether it might be singles, divorced people, gay people, or whatever. The interesting thing about diversity is that most customers don't like to be with people who are not like themselves.

And there's a problem of what I call market orchestration. How do you orchestrate very diverse groups and have a successful institution? That alone puts pressure on trying to define your market. It's not everyone; but it's more than one group. The church needs well-defined groups who are looking for one or more particular satisfactions.

PETER DRUCKER: So the mission may well be universal. And yet to be successful, the institution has to think through its strategy and focus on the main target groups in marketing and delivering its service. The same thing is true for fund-raising, isn't it?

PHILIP KOTLER: Fund-raising requires careful identification of the appropriate sources of funds and the giving motives. Why does that donor give money? To whom does the donor give money? And so I would again say that consumer research is important in the process of trying to direct your efforts.

PETER DRUCKER: And to what extent do you then have to mold what you are, do what you can for the market? In the Church, for instance, there is a community of older people that's your prime community, but they really want a very different church from the one that attracts the singles; and so each church would then have to change what it does to serve its high potential market.

PHILIP KOTLER: The Church may establish different services and different ministries for its different groups. For example, it could have an early morning service for one group and a later morning service for another. I think the solution there is probably to have different leaders and lay ministries serving the different groups.

PETER DRUCKER: But you don't seem to believe that, what in reaching markets one would call "boutiques," are very successful for non-profit institutions?

PHILIP KOTLER: Translate boutiques into niches! I believe that some organizations should definitely go the route of niching versus mass production. For example, take theater groups. The city of

Chicago has over 120 performing art theatrical groups. What some of these theater groups have done is to niche into a certain class of performances. There's one that does Shakespeare only, another that does the classics in general, another that does only plays written in the last ten years. The question is, Do you want to satisfy one type of audience deeply or do you want to satisfy a number of audiences more superficially?

PETER DRUCKER: You know, I've done a fair amount of work with museums, and the really successful ones are building niches very strongly. The universal general museums of the nineteenth century, of which the Metropolitan in New York is still the leading American example, are becoming . . . well, old-fashioned. They have no real clientele. But museums can be too narrow. We have a wonderful museum here in Los Angeles of the American Indian and it's too narrow. But I think we see more and more niching, even in hospitals, where the community hospital is in a sense giving way to boutiques; there's a free-standing surgical unit, and there is a specialty hospital. I think we need product differentiation in the non-profit institution as much as we need it in business.

PHILIP KOTLER: I have to agree with you. But that does pose a problem for the nineteenth-century-type institutions. Do they break themselves up? Should General Motors split itself into five different companies? The "supermarkets," basically, see themselves as having a marketing problem. The way the Art Institute in Chicago has handled this is by forming groups of loyal donors and supporters around different art forms. There's a modern art group that has a meeting once a month, and they always have a lecturer or they see some new developments in modern art. There's another group that's Ancient Greece and Rome. And so it is possible in a major museum still to form interest groups. You know, small is beautiful. How do you help your customers to identify with something that is as ungraspable and huge as a major museum?

PETER DRUCKER: Well, we have that problem, I think, in a good many institutions. We have it in the church and the synagogue. And a good many of my friends in religious institutions have to grapple with being identifiable and yet at the same time not becoming separatist. We have it, I think, at its most extreme in the university where, if you look for the institution that has done the best marketing job, it's the fundamentalist college. Precisely because it is a boutique, it doesn't try to do anything but a very narrow specialty. And, on the other hand, the research university has done quite well. But the comprehensive universities that did so well in the fifties and sixties are beginning to, I would say, lose character in the public mind. That explains why the good liberal arts college, which we all thought was going to be in severe trouble fifteen years ago when student populations began to go down, is doing so very well. It isn't so small; twenty-five hundred students isn't tiny. But the kid can get his or her arms around it, and it has a personality, whereas the University of Minnesota or UCLA are very hard to describe. I think we will see a good deal of—not niche marketing in the non-profit section but product identification, as you would call it in a business. The market, very largely, will determine the character of the institution and the character of the product.

Why does the non-profit institution have to be interested in marketing and have to engage in marketing? Is it to be sure that it really fulfills the need? Will it satisfy the customer? Is it to know what it should focus its energies on? What are the real reasons for doing marketing for a non-profit institution?

PHILIP KOTLER: Marketing really is spurred by the presence and the increase in competition that the institution faces in a way that it never faced before. Most organizations don't get interested in marketing when they are comfortable. Suddenly they find that they don't understand their customers very well, and their customers are leaving that church, or they're not signing up for that college, or coming to that hospital. And these institutions become aware of a competitive situation.

How do you deal with a competitive situation? Well, one way some early hospitals dealt with it was to pray that the world hadn't changed and that they would just survive. Now, prayer may have its role to play, but it is not the answer. The normal answer is that maybe there's something in this thing called marketing that will help us understand why customers chose to be with us in the first place and why they're not choosing to be with us anymore.

PETER DRUCKER: Philip, it's an old theological axiom that prayer is no substitute for right action. And that's what you're telling us. Who then should really do the marketing job in the non-profit institution?

PHILIP KOTLER: The chief executive officer should, of course, be the chief marketing officer. Marketing doesn't get anywhere in an organization without the head of the organization getting interested in it, understanding it, and wishing to disseminate its logic and wisdom to the staff and people connected with the institution. Still, the CEO can't do the marketing. The work has to be delegated to someone who is skilled in handling marketing. Most institutions appoint a director of marketing or a vice-president of marketing. Those are, for example, the titles you will see in hospitals. There's a difference, of course. The director of marketing is seen as a person who has "skills will travel," and not someone who is in a policy-making or policy-influencing position. That's why I favor a vice-president of marketing position, because that person really should sit with all the other officers as they try to visualize what the future of their institution will be.

PETER DRUCKER: And how can we tell whether marketing in a non-profit institution, this church and this synagogue, this hospital and this college, is making a genuine contribution?

PHILIP KOTLER: Marketing is supposed to do the following. It is supposed to build up what I call share of mind and share of heart for the organization. At any point in time, the institution or organization has a certain level of awareness in its target market and a certain amount of favorable attitude. A good marketing program will build up more awareness and more loyalty or bonding with the public you are trying to serve. So, one way to measure the contribution of marketing is to see whether more people know about our church, and more people like our church, or whatever the institution might be. There's a cost side. Budgets have to be developed for the work that must take place. And it is very hard to gauge the impact of marketing without setting objectives. If an institution said, We would like to go from 30 percent of the target market knowing about us and 80 percent of those who know about us, liking us, to 90 percent of those who know about us liking us, then that's measurable. It's measurable through normal marketing research. So, the key to knowing whether marketing is working is to set objectives and then to see if marketing has helped the organization to realize them.

PETER DRUCKER: And the more specific the objectives, the more likely to be productive?

PHILIP KOTLER: Absolutely. The problem has arisen in hospitals lately that the hospitals have used their budgets for advertising purposes. They have spent big dollars trying to communicate to their communities that they are a friendly hospital, they are a "caring" hospital, and so on. And they are all wondering now whether those ads have really established in the minds of the community an identity for that hospital and a preference for the hospital. Some CEOs are disturbed about the results; they don't see enough net gain.

My analysis is that these hospitals have often put their budgets to the wrong use. They've gone into heavy advertising before they had a character to their hospital. Before they had a true patient focus in their hospital. And they haven't really gone into marketing in the right order. The order being: first, do some customer research to understand the market you want to serve and its needs. Second, develop segmentation and be aware of different groups that you're going to be interacting with. Third, develop policies, practices, and programs that are targeted to satisfy those groups. And then the last step is to communicate these programs. Too many hospitals and other non-profit organizations go right into advertising before they've gone into the other three steps, and that's really doing things backwards.

PETER DRUCKER: And to talk hospitals, I know far too many that would resist to the bitter end the kind of communication their market research shows them the public wants, which is how many of the people who come in to have a hip replacement can walk after six months. Because not everybody does. If we say 98 percent can walk, that means 2 percent can't. And then we gloss it over by saying, "We love you." Patients facing major surgery have other worries than being loved. What you're saying is that you have to start out with knowing what the customers really consider value, what is important, before you communicate, rather than with telling the things you believe should be important to the customer. That's the key to effective marketing.

PHILIP KOTLER: It is. I've often said that non-profit organizations that have no marketing, or little marketing, will probably take five to ten years to really install effective marketing procedures and programs if they're fully committed to installing them. And mind you, many organizations give up after one or two years, especially if the early results are so good that they think they are already there. It takes five to ten years because marketing is more than a department, it's really everyone in the organization pursuing one goal and that is to satisfy the customer, to serve the customer. So, getting the other departments in the museum, getting the curators to understand that, getting the janitorial and the maintenance staff and the guards to understand that—it's hard and takes time.

PETER DRUCKER: What you are saying is that marketing in an organization is everybody's business, certainly everybody who has anything to do with the customer. So you are talking not about a function—though there is specific work—you're talking of a basic commitment. In talking of marketing in the non-profit organization, you are talking of the basic action that results in an organization that is both dedicated and positioned to satisfy its basic purpose.

PHILIP KOTLER: Exactly. Marketing in a non-profit organization becomes effective when the organization is very clear about what it wants to accomplish, has motivated everyone in the organization to agree to that goal and to see the worthwhileness of that goal, and when the organization has taken the steps to implement this vision in a way which is cost-effective, in a way which brings about that result.

PETER DRUCKER: So, would you agree that marketing is the work—and it is work—that brings the needs and wants and values of the customer into conformity with the product and values and behavior of the supplier, of the institution?

PHILIP KOTLER: Marketing is a way to harmonize the needs and wants of the outside world with the purposes and the resources and the objectives of the institution.

4

Building
the Donor Constituency

Interview with Dudley Hafner *

PETER DRUCKER: What we used to call fund-raising, we now call fund development. Is this pure rhetoric, Dudley?

DUDLEY HAFNER: For some, it might be pure rhetoric, but for others it's recognizing that your true potential for growth and development is the donor, is someone you want to cultivate and bring along in your program. Not simply someone to collect this year's contribution from.

PETER DRUCKER: Does that apply only to national organizations such as yours? Or would you say the same thing about the United Way or the local church or the local hospital?

DUDLEY HAFNER: It applies to all of the non-profit organizations. One of the things that helps an organization move forward is to have a broad, sound, solid advocacy base. One of the places to develop that is within your giver group. You need those advocates.

*Dudley Hafner is executive vice-president and CEO of the American Heart Association.

PETER DRUCKER: But also, of course, it must greatly reduce acquisition cost; the cost of getting the money, when you have a donor base that is already sold. You don't have to sell every year. Is that right?

DUDLEY HAFNER: That's correct. It's just much more efficient to organize with the notion that you are going to have a long-term relationship with your donors, that you're going to help them increase their support to the organization. But from an effectiveness standpoint, it also makes a lot of sense because for a non-profit organization to be really successful, you have to have a lot of people caring about how it does. You want that donor to take ownership in your program.

PETER DRUCKER: What are some of the tools you use in your sixteen hundred local organizations? That's where you raise most of your money, isn't it?

DUDLEY HAFNER: Ninety-nine percent of it is raised at the community level. First of all, what you want to do is acquaint donors with what you are as an organization, what you are trying to get accomplished, so they can identify with your goals.

PETER DRUCKER: For this, Dudley, you have to have a very clear mission, don't you?

DUDLEY HAFNER: You have to have a very clear mission and very clear goals. Our goals relate directly to our mission, which is the prevention of premature death and disability from cardiovascular disease and stroke. The kinds of goals that relate to that would be the number of people that we convince to stop smoking or using tobacco or not start in the first place, or people that change their dietary habits, or biomedical research we want to fund. It all has to be tied back to the benefits to the community.

PETER DRUCKER: Let's say you come to me. What would you say to me before I put my check into the envelope?

DUDLEY HAFNER: We present a case for support which spells out the magnitude of the challenge, what we propose to do about it, how realistic it is to achieve that challenge, and how your gift can make a difference. In cultivating you, we would do that perhaps in a series of mailings throughout the year. If we were really cultivating you, we might ask you to get involved in some of our activities.

PETER DRUCKER: Such as ringing the doorbells in my neighborhood?

DUDLEY HAFNER: Do that, or help us give blood-pressure screening programs. Cultivating you as a donor means giving you a chance to make a difference in what it is we're trying to achieve

PETER DRUCKER: And you have basic goals. First you have to get people to start giving, and then you have long-term goals for making them what I would call members in terms of their commitment, in terms of their being really not outside donors, but people concerned with the success of the organization.

DUDLEY HAFNER: Development means bringing the donors along, raising their sights in terms of how they can support you, giving them ownership in the outcome of your organization. That takes a long-term strategy rather than putting together an annual campaign to go out and collect money.

PETER DRUCKER: You know I've heard it said the American Heart Association or the Cancer people have it easy because the donors really give to themselves. We in the international field or in universities can't appeal to the donor's self-interest. Is that a meaningful argument?

DUDLEY HAFNER: People in the non-profit health community look at academia and the colleges and say, Oh, but they are blessed with those large corporate foundation-giver types that we wish we could have. Most of our gifts are in the five-dollar range.

We all have our special groups of interest and our challenge is to expand those groups of interest.

PETER DRUCKER: I think you said one of the most important things, and I only wish more people would listen to it: You have to think through to whom you make sense, basically.

DUDLEY HAFNER: That's exactly right. Then appeal to them in a very forceful, forthright manner.

PETER DRUCKER: You know, Dudley, to me one of the most amazing things is how few people in the United States understand the importance and the uniqueness of the things you are talking about. My European friends always point out how low the taxation rate is in the United States. I say, you are mistaken because we voluntarily cough up another 10 percent of GNP for things which in Europe are either not done at all, like your work, or run by the government with the individual having absolutely no say in where the money is to be spent. That's a point the public does not understand. Would you agree?

DUDLEY HAFNER: I agree. There's a couple of things about this that are very, very important to me personally. First of all, campaigns such as the American Heart Association or the Salvation Army or the Girl Scouts let people get involved, and that becomes important because they do become advocates. The other thing I think that is unique about these United States is the fact that charitable giving is as much a force in the freedom of democracy as the right of assemblage or the right of vote or the right of free press. It's another way of expressing ourselves very, very forcefully. Someone who pays taxes does not think of himself or herself as getting involved in the welfare program. But if they become involved in a Salvation Army activity or the Visiting Nurses program, they *are* involved. They are involved spiritually and they are involved monetarily. That makes a difference.

PETER DRUCKER: We talk blithely about volunteerism without explaining what we mean. But to come back to creating that constituency of yours or the constituency for the local church or the local hospital or the local Girl Scout Council or Boy Scout Council, or for any national organization: What kind of materials do you supply? What kind of tools do you supply to the people who come to me and say, "Will you collect in your neighborhood and here is the kit?" How do you work that out?

DUDLEY HAFNER: We have a prescribed structure that we offer to the local leadership. We have job descriptions. We have a way for them to formulate goals for now and five years out. And then we have the materials that support each one of those elements of the fund-raising.

Those materials are created after we look at the various segments of our giver groups. We know through market research the preferences of, say, a family that's in their fifties and has an income of a certain level, and a family that's in their thirties and a different income. So, with a handle on that kind of information about values and aspirations, we can develop the materials that will deliver the same message, but in a different way so that it catches the attention of the individual.

PETER DRUCKER: You said two things I heard loud and clear. One is, you said market research. You go to great lengths to study the market and to focus your message on what in marketing we would call the values of the potential customers. And the second thing, you have very clear goals for a marketing campaign in which you market the American Heart Association to potential investors, to people willing to commit themselves, if only in the beginning to a token donation just to get rid of the collector. And I've had lots of people who said, "Tell me how much you want so I can go back to the TV set." I'm actually quoting. But next year the same person very often says, "That literature you left was very interesting." That's when I have learned to say, "Last year you gave ten dollars; how about twenty-five this year?" And half the time I get it.

DUDLEY HAFNER: Peter, you are a fine fund-raiser because you are dealing with what is essential to a successful campaign—that every donor is very, very valuable to you. You may acquire that donor going door-to-door. And it may be a dollar gift to get rid of you so they can go back to the TV set. But an organization that is concerned about its future will keep track of that dollar, and next year they'll go back and they'll encourage that to be $2, or $5, or $10, if it appears that individual could be giving at the level.

Every donor becomes very, very precious.

PETER DRUCKER: But you know, the most valuable training as a door-to-door fund-raiser I got was not from you but from another organization which said, "Don't go Sunday afternoon when the professional football games are on. Then you can't get them away from the television even for those two bucks." And I found out they are right. I'm impressed when I go around by the difference, by the way, between the support I get from your organization as against the weak support from another where I'm not able to answer the questions I get. The difference between your enabling your field salesman to be an effective spokesman for the organization and the ones where all I can do is appeal to "You know how many babies are dying." That gets money out if there has been a horror story on TV or in the newspaper headlines yesterday, but not otherwise.

DUDLEY HAFNER: For long-term growth of an organization, you have to appeal to the rational in the individual as well as the emotional part of the individual. In building local campaigns, you have to think of the person who does door-to-door, who is treated as a salesperson by a potential contributor. See it as an opportunity to educate those potential donors about what they can do for themselves personally, if it's a disease. What they can do in terms of the overall mission, in terms of concerns, plus their gift. And if you don't use that opportunity to do that, you are not building on your greatest opportunity to create a long-term strategy.

PETER DRUCKER: And despite all the tremendous competition for funds—no day goes by without three, four, five appeals—the amount of money you are able to raise for your mission is going up, or holding steady at least?

DUDLEY HAFNER: We are well ahead of inflation. Let me say something, Peter, about competition in this area. The American Heart Association or the American Lung Association cannot afford to create a strategy, in my opinion, that will cause one of them to do better at the expense of another non-profit health-care organization. So, what we have to do is to figure out how to get new monies that have not been previously given, rather than have someone transfer their allegiance from one non-profit program to another one. To have a long-term really positive impact on the good that the non-profit organizations are trying to do.

PETER DRUCKER: I never heard this before and I am impressed. It seems to me almost to be the opposite of what I hear all the time when that college, or that church, or that hospital, or that national organization says, "We want people who give to nobody but us." May we go back to something we began to talk about and then left aside—your market research? Tell us a little more about it.

DUDLEY HAFNER: We do market research because we feel a commitment to the 2.5 million volunteers who go out as our ambassadors. We give them the best possible materials. We're giving them things we know will work.

PETER DRUCKER: What kind of knowledge about the market is relevant?

DUDLEY HAFNER: What kinds of prior experience in that person's life will cause them to be more responsive? What are they dealing with today that is the button you want to press in terms of having them see you as a unique organization? You have to rise above all that clutter of information out there about what to buy, what to do with your leisure time, and what charitable organizations, volunteer organizations you support. That information makes us more effective with our message and building our case for support. Our volunteers are more focused.

PETER DRUCKER: You know, every fall I get the brochure from a local organization, and it says at this level of income you should give so much, and I've always wondered whether this is productive or counterproductive?

DUDLEY HAFNER: What we've found in asking for a specific gift is that it dramatically improves the return in our campaign. I would say that organizations running annual campaigns without asking for specific gifts could, with the same effort, probably increase their income by as much as 25 percent by asking for a specific gift.

PETER DRUCKER: So I was wrong as usual.

DUDLEY HAFNER: Let me tell you what I think is at play here. People who find the appeal sets its sights a little bit too high are not offended; they're usually flattered. For the individuals who are being asked to give less than they had in mind, we find they tend to go ahead and give what they had in mind anyway, and you can build from there. Once you've given a gift that is suggested, you fall into a category that the non-profits should pay special attention to: the long-term strategy of upgrading that gift.

PETER DRUCKER: How do you do that? Do you pick out the people who give more than suggested as your first target of opportunity?

DUDLEY HAFNER: That's one way. Then you also have a strategy to increase the size of the gift you ask for each year from those people who have given the suggested amount. And I'm not talking about in a crass way; I'm talking about in a way that just gently nudges them to a higher level. I've been involved in local campaigns in which we didn't know individuals. We suggested a certain amount. And those were the gifts that came in.

PETER DRUCKER: Do you single these people out by way of giving them more information, or how do you build the relationship?

DUDLEY HAFNER: You classify that individual by the kind of follow-up, which can range from a personalized thank-you letter, to inviting them to specific activities, to sending an annual report, showing them what you're planning to do with the money or how the money has helped.

PETER DRUCKER: The constant emphasis is on the mission, basically, to upgrade your potential high-yield donors.

DUDLEY HAFNER: That's exactly right.

PETER DRUCKER: So, your market research tries to identify two things, to use technical terms: both market segmentation and market value expectations. Are the market segmentations strong?

DUDLEY HAFNER: Our research says we deal with forty-one different discrete markets.

PETER DRUCKER: Give me a couple of examples.

DUDLEY HAFNER: Someone age fifty and making $40,000 a year wants to be solicited differently from someone age thirty, with children at home and an income of $25,000.

PETER DRUCKER: Are there any groups that simply aren't customers at all?

DUDLEY HAFNER: For the Heart Association, I don't think so. Although if you're a fund-raiser as such, you might say there are certain areas that you don't want to put a lot of time in because your contribution base is not going to grow that much. But there's a piece of me that says that this is more than just raising monies for this organization. It's an educational opportunity and it's an opportunity to let somebody be involved. If that's for a quarter or a dollar, it's still worth it.

You cannot build your long-term growth strategies, income strategies, on that philosophy, however. It has to be built on cultivating the larger donors and raising their sights.

PETER DRUCKER: Well, you have to go where the money is to get it, and that is very important. But you also look upon fund development as an educational campaign, not just to get money but to strengthen the objectives of the American Heart Association.

DUDLEY HAFNER: Absolutely, and that's part of the justification for having a broad-based annual campaign. You have to have a strategy for your fund development and know what you expect out of the various strategies, what your return expectations are. Then you measure your success against that. With the larger givers, you have one strategy and one expectation. Smaller givers, another strategy and another expectation.

PETER DRUCKER: You know, strategy is a very popular word just now, but what precisely do you mean by "strategy"?

DUDLEY HAFNER: For me, it's how we use our resources to get the attention of that individual to do what it is we hope he or she will do.

PETER DRUCKER: And it is always focused on an individual in the end?

DUDLEY HAFNER: It's always focused on an individual.

PETER DRUCKER: Let's say, you single out one of those forty-one markets of yours by age and income, and maybe you have urban or suburban or rural. How do you develop what you call a strategy?

DUDLEY HAFNER: If we're going after people who are age fifty, which is a high-risk age, we want to show these individuals how they can reduce their risk of heart attack. How research or education is going to have immediate feedback, because that's their interest. So, your strategy is to provide something they can relate to—and give to at the same time.

PETER DRUCKER: Do you supply your fund-raisers, those local volunteers, with information about the potential donors before they go to them? Or do you just say, If this is a fifty-year-old man you use strategy A, and a twenty-five-year-old woman you use strategy B?

DUDLEY HAFNER: You will receive materials based on the neighborhood in which you live. There is an awful lot of very good data now that we can roll out on any community in this country and say that within this section of the community these are the materials that will be of most interest to people you call on. These are general statements and certainly there will be exceptions. Put your materials together. Then have a volunteer in that neighborhood go door-to-door with that material, call on the individuals, and make a much greater impact.

What's emerging for the future, and I hope for the non-profit, is not organizing in the traditional fashion—special gifts, special events—but around value groups. Make each one of those value groups an identified market, with their own materials, their own strategies, their own support system. The primary factors in the value groups, of course, will be age and then income. After that, there's a whole host of other things that you can use, but I think that for the day-to-day operation of most of our non-profits, those additional values are not going to make that much difference.

PETER DRUCKER: If you were to pick out one or two factors that are crucial to fund development and fund-raising, whether it's a national organization or local one, or whether it's as big as you are or it's that neighborhood shelter for battered women, what would you pick out?

DUDLEY HAFNER: I'd pick out the care and treatment and cultivation of the donor. That's number one. The second thing I would do is ask for a gift that is in relationship to the individual's ability to give. Those two things will give you long-term, stable growth. It will give you a broad-based advocacy, and I think those are the two most important parts.

PETER DRUCKER: You wouldn't put identification of potential donors that high?

DUDLEY HAFNER: Donor acquisition is very, very critical. But I'm often disappointed to find that an organization has made a considerable investment in donor acquisition and then failed to put that donor into their files in such a way that they can continue to cultivate him or her. So, the initial investment is never truly realized to its fullest potential.

PETER DRUCKER: Well, let me try to pull out what I think are the central points. You have told us, first, of the central importance of the clear mission, and the importance of knowing your market, not just in generalities, but in fine detail. And then of enabling those volunteers of yours to do a decent job by giving them the tools that make it almost certain that they can succeed. And finally what I heard you say loud and clear is that you don't appeal to the heart alone, and you don't appeal to the head alone. You have to have a very rational case, but you also must appeal to our sense of responsibility for our brethren.

DUDLEY HAFNER: Exactly, you must do both if you are going to have a long-term growth in your development program.

PETER DRUCKER: Dudley, the one area we have not really talked much about is volunteers. Do you really need the volunteers, or can you do that today with the computer and TV? I see so much telemarketing in fund-raising efforts by non-profit organizations.

DUDLEY HAFNER: I'm really glad that you brought us back to that point because I think that many organizations may be facing a crisis in their future—I hope that they're aware of that. To answer your question, do we need volunteers to raise money next year? Technology has given us the means to go out and probably do a pretty good job of raising money through the computer, through mail drops or telemarketing that leave out the volunteer. But that would be a tragic mistake because in the process you've also lost the constituency, you've lost the volunteer base, you've lost the course of strength and growth in the organization. I see technology as a way of *helping* the volunteers do a more effective job; I do not see it as a replacement for a volunteer. And I think any organization that makes that connection—makes the decision that it's easier to raise money without involving the volunteers—will have made a fatal mistake.

PETER DRUCKER: Let me try again to sum up. I think the strongest thing you said just now to me is that fund development is *people* development. Both when you talk of donors and when you talk of volunteers. You are building a constituency. You're building understanding, you're building support. You're building satisfaction, human satisfaction in the process. That is the way to create the support base you need to do your job. But it's also the way you use your job to enrich the community and every participant. And it's based on clear mission, on extensive and detailed knowledge of the market, of making demands on both your volunteers and your donors, but also on feedback from your performance, which, I think, is something on which a good many non-profit organizations are pretty weak. You never hear from them what the results are. And I think that what you said may be even more important for the purely local and small organization, precisely because on the local scene you have a lot of well-meaning people, but very often you have no sense of direction. You have a need, but no message. I hope what you told us will be heard and applied, particularly by the local organization, where the need is so great and where good intentions just aren't good enough.

5

Summary:
The Action Implications

Strategy converts a non-profit institution's mission and objectives into performance. Despite its importance, however, many non-profits tend to slight strategy. It seems so obvious to most of them that they are satisfying a need, so clear that everybody who has that need must want the service the non-profit institution has to offer. One central problem is that too many non-profit managers confuse strategy with a selling effort. Strategy *ends* with selling efforts. It begins with knowing the market—who the customer is, who the customer should be, who the customer might be. The whole point of strategy is not to look at recipients as people who receive bounty, to whom the non-profit does good. They are customers who have to be satisfied. The non-profit institution needs a marketing strategy that integrates the customer and the mission.

An effective non-profit institution also needs strategies to improve all the time and to innovate. The two overlap. Nobody can ever quite say where an improvement ends and an innovation begins. When Frances Hesselbein and the Girl Scouts introduced their new service for five-year-olds, the Daisy Scouts, that was, in one way, just old-fashioned Girl Scouting. In another way it was a drastic innovation.

And then the non-profit institution needs a strategy to build its donor base. It needs to develop a donor constituency.

All three of these strategies begin with research and research and more research. They require organized attempts to find out who the customer is, what is of value to the customer, how the customer buys. You don't start out with your product but with the end, which is a satisfied customer.

The most important person to research is the individual who *should* be the customer, the people who are believers but who have stopped going to church. Traditionally, businesses have researched their own customers and know, or try to know, as much as possible about them. But even if you have market leadership, non-customers always outnumber customers. The most important knowledge is the *potential customer.* The customer who really needs the service, wants the service, but not in the way in which it is available today. The typical college or university, after twenty years of an enormous number of young people reaching college age, is only now accepting the fact that it has to market the college to high school counselors, to prospective students, to their parents. Despite a sharp drop in the total number of applicants, colleges that do market effectively have more applications than they can possibly admit.

You would imagine that people would be only too eager for services aimed at helping them prevent a heart attack, or recover from it. Yes, they are, but only if the service fits them—their age, their weight. They manage their own life and their own health.

This understanding of the importance of strategy is particularly crucial to non-profit managers when it comes to their donors.

The typical non-profit institution still goes around telling donors, "Here is the need." But the ones that get results—the ones that attract and build a fund constituency—say, "This is what *you* need. These are the results. This is what we do for you." They look upon the donor as a customer. This is the essence of a strategy: it always starts out with the other side. Even thousands of years ago, the beginning of wisdom in military strategy was to start out with the enemy and not with your own troops.

The next step in non-profit strategy (as in military strategy) is the training of your own people. Everyone in the hospital must be patient-conscious. That's a training job—not just preaching. It isn't attitude, it's behavior. In fact, we have learned that attitude training is not very effective. The way to train people is behavior-ally: *This is what you do*. With that kind of specific training, even hospital workers who are very far away from the customer—the billing office, the janitorial workers—do things that satisfy the customers: the physicians, the patients.

In non-profit management, training doesn't apply only to the employees; training volunteers may be even more essential, espe-cially in an organization in which volunteers are the interface with the customers, with the public.

When it comes to introducing something new, when it comes to innovation, non-profit strategy requires careful thought and plan-ning: where to start and with whom. Start with people who want the new to succeed. Don't try to have everybody in the organiza-tion run with the new first. That route always gets into trouble.

Look for a target of opportunity, for somebody in the organiza-tion who wants the new, who is convinced of it, who is committed to it. The strategy in innovation is to think through this process at the start, so that you can identify somebody willing to work hard at making the new successful, and somebody whose success then becomes a multiplier in the organization.

The worst thing in strategy is to introduce something with great fanfare and great hope that it is going to change the world, and five years later say, "Well, it's doing all right. It's a little specialty." That's failure. That's misallocation of resources.

Knowing the customer also enables the non-profit organiza-tion—whether it's the church, the synagogue, the Scouts, a hospi-tal, a college—to know what results to expect. It is important to define goals and know what realistically should work. What are we trying to do? This college is trying to get in so many applicants, of what quality, so that it can maintain its size and quality. Then one can feed back from results. Then one can say, "Well, we are doing quite well here, but not really well over there. Let's put in a little more effort." Or, "We need a stronger person in charge." Or, "We need to offer something additional that will bring in the kind of students we need."

Strategy also demands that the non-profit institution organize itself to abandon what no longer works, what no longer contributes, what no longer serves. A church must get out of the singles ministry if it doesn't have the right person to run it and cannot guarantee a quality service. The American Heart Association must be willing to play down older people as potential donors because to people over seventy-five or eighty, death by heart attack is not the worst of all possible ways to go. That's abandonment. If you don't build it in, you'll soon overload your organization and put good resources where the results don't follow.

The question always before the non-profit executive is: What should our service do for the customer that is of importance to that customer? Then think through how the service should be structured, be offered, be staffed. End up with nuts and bolts: What to do, when to do, where to do. And most importantly, who is to do it?

Strategy begins with the mission. It leads to a work plan. It ends with the right tools—a kit, say for volunteers, which tells them who to call on, what to say, and how much money to get. Without that kit, there is no strategy.

The last thing to say about strategy is that it exploits opportunity, the right moment. Greek theologians called it *Kairos,* the point when the new is received. Most of the needs non-profit institutions fill are likely to be there forever in one form or another; they are parts of the human condition. But the need presents itself in a specific form, and it is the function of research to find out, at this time, what that form is. Especially for the ones who should be customers but aren't because the service is not available in a form that serves them. Ask: "Is this something that fits our strengths? Can we develop the service that satisfies?" Then comes that third element, the right moment to seize the opportunity by the forelock, to run with success.

Strategy commits the non-profit executive and the organization to action. Its essence is action—putting together mission, objectives, the market—and the right moment. The tests of strategy are results. It begins with needs and ends with satisfactions. For this you need to know what the satisfactions should be for *your* customers: the parishioners in your church, the sick in your hospital, the boys and girls in your Scout troops, and the volunteers who lead them. What is really meaningful to them? Non-profit people must respect their customers and their donors enough to listen to *their* values and understand *their* satisfactions. They do not impose the executive's or the organization's own views and egos on those they serve.

PART THREE

Managing for Performance

how to define it;
how to measure it

1

What Is the Bottom Line When There Is No "Bottom Line"?

Non-profit institutions tend not to give priority to performance and results. Yet performance and results are far more important—and far more difficult to measure and control—in the non-profit institution than in a business.

In a business, there is a financial bottom line. Profit and loss are not enough by themselves to judge performance, but at least they are something concrete. Whether business executives like it or not, profit certainly *will* be used to measure their performance. When non-profit executives, however, face a risk-taking decision, they must first think through the desired results—*before* the means of measuring performance and results can be determined. For each non-profit institution, the executive who leads effectively must first answer the question, How is performance for this institution to be defined? In a hospital emergency room, for instance, is performance how fast the staff sees people who come in? Is it the number of heart-attack victims who pull through the first few hours after they arrive? What is the performance of a church? One may look strictly at attendance; but there is also the impact on the community. Both are perfectly respectable ways to measure performance, yet each leads to a very different way of running the church. An organization to tackle AIDS does not have to worry about the need for its efforts. But it must be clear whether its performance is to be measured by success in prevention of the disease or in taking care of AIDS patients. If the aim is prevention, the organization has to create its own customers, the people who do not have AIDS and tend to believe that AIDS is what other people contract.

It is not enough for non-profits to say: We serve a need. The really good ones create a *want*. Museums, for instance, used to see themselves as cultural custodians. Their administrators believed in keeping art in and people out. Most museums today work hard to create customers for taste, for beauty, and for inspiration. They see themselves as educational institutions. The Cleveland Museum became one of the world's great museums not only because it had a director who was a whiz at finding great objects; he was equally adept at making patrons out of "casuals," people who just dropped in to spend an idle hour out of the rain. He used terms such as "repeat sales" to measure the performance of his institution. As he saw it, building the percentage of repeat sales built a clientele, built a community institution rather than a comfort station.

As non-profit executives begin to define the performance that makes the mission of their institution operational, two common temptations have to be resisted. First: recklessness. It's so easy to say that the cause is everything, and if people don't want to support it, too bad for them. Performance means concentrating *available* resources where the results are. It does not mean making promises you can't live up to.

But equally dangerous is the opposite—to go for the easy results rather than for results that further the mission. Avoid overemphasis on the things the institution can easily get money for, the popular issues, the easy things. Universities, for instance, often are under great pressure to accept money for a chair that administration and faculty feel actually detracts from the school's mission (we call them "Mickey Mouse chairs").

Lately, I have been worrying over a similar problem with an art museum. A patron is offering to give the museum an outstanding collection, but under conditions that would impair the museum's main mission. One possible response is to be virtuous and say no. The other is to be dishonest and sign on the dotted line, knowing that the donor won't live forever; after all, you are being dishonest in a good cause. But if we accept, we'll pay a heavy price. The whole organization will become cynical. Yet the temptation is great. If we say no, another less scrupulous museum will get that fine collection.

Both temptations have the same root: the non-profit doesn't get *paid* for performance. Even if it can charge fees for its service—the entrance fee the museum charges, for instance, or the money a well-run museum shop now earns—the non-profit cannot generally generate more than a fraction of the funds it needs to operate. In a business, performance is what the customer is willing to *pay* for. The non-profit does not get paid for performance. But it does not get money for good intentions, either.

PLANNING FOR PERFORMANCE

Performance in the non-profit institution must be *planned*. And this starts out with the mission. Non-profits fail to perform unless they start out with their mission. For the mission defines what results are in this particular non-profit institution.

And then one asks: Who are our constituencies, and what are the results for each of them?

One of the basic differences between businesses and non-profits is that non-profits always have a multitude of constituencies. It used to be that a business could plan in terms of one constituency, the customers and their satisfaction—the Japanese still do. Everybody else—employees, the community, the environment, maybe even the shareholders—were restraints. That this has changed for American business, and quite drastically, is the reason why many business executives feel the world is coming to an end. But in the non-profit institution there have always been a multitude of groups, each with a veto power. A school principal has to satisfy teachers, the school board, the taxpayers, parents, and, in a high school, the students themselves. Five constituencies, each of which sees the school differently. Each of them is essential, and each has its own objectives. Each of them has to be satisfied at least to the point where they don't fire the principal, go on strike, or rebel.

Thirty years ago, community hospitals were run basically for the physicians. Physicians were the buyers. The physician said, "I'm going to put you into this hospital," and it did not occur to the patient to say no. That's now gone. And one of the reasons hospital management is becoming so difficult is that third-party payers, the companies who pay for their employees, have now become a constituency that has to be satisfied, both medically and economically. Uncle Sam, too, has become a very powerful constituent since about two fifths of the revenue of the typical community hospital comes from Medicare. The new health providers, the health maintenance organizations (HMOs), have become constituents. And the hospital's personnel have become far more important, not because they demand more, but because so many more are now highly trained, professional people.

The success of the growing pastoral churches largely depends on their realizing that the needs of young people, young married couples, singles, and older people are different. The church has to set a performance goal with respect to each group and use competent individuals who can deliver performance. One of the country's largest and most successful churches gave up its ministry to the singles because it could not find a truly competent assistant pastor to run it.

The first—but also the toughest—task of the non-profit executive is to get all of these constituencies to agree on what the *long-term* goals of the institution are. Building around the long term is the only way to integrate all these interests.

If you focus on short-term results, they will all jump in different directions. You'll have a flea circus—as I discovered during my own dismal failure some forty years ago as an executive in an academic institution. My own thinking has always been long term. But I thought I'd win friends and influence people by giving them some short-term goodies. What I learned was that unless you integrate the vision of all constituencies into the long-range goal, you will soon lose support, lose credibility, and lose respect. After

I'd been beaten to a pulp, I began to look at non-profit executives who did successfully what I had unsuccessfully tried to do. I soon learned that they start out by defining the fundamental change that the non-profit institution wants to make in society and in human beings; then they project that goal onto the concerns of each of the institution's constituencies.

This kind of planning is quite different from what business people usually mean by the term. To formulate the plan successfully, non-profit executives think through the concerns of each of the institution's constituencies. They try to understand what is really important to an elected school board, to the faculty of the school, to the parents of the students. Long-term concerns must be identified—not short-term concerns such as the parents who worry whether their Marilyn will get into the college of her choice. But for a school to be good enough so that its students have a choice where they go to college is a legitimate long-range goal for both constituencies, parents and their high school children. Integrating constituency goals into the institution's mission is almost an architectural process, a structural process. It's not too difficult to do once it's understood; but it's hard work.

MORAL VS. ECONOMIC CAUSES

The discipline of thinking through what results will be demanded of the non-profit institution can protect it from squandering resources because of confusion between moral and economic causes.

Non-profit institutions generally find it almost impossible to abandon anything. Everything they do is "the Lord's work" or "a good cause." But non-profits have to distinguish between moral causes and economic causes. A moral cause is an absolute good. Preachers have been thundering against fornication for five thousand years. Results, alas, have been nil, but that only proves how deeply entrenched evil is. The absence of results indicates only that efforts have to be increased. This is the essence of a moral cause.

In an economic cause, one asks: Is this the best application of our scarce resources? There is so much work to be done. Let's put our resources where the results are. We cannot afford to be righteous and continue this project where we seem to be unable to achieve the results we've set for ourselves.

To believe that whatever we do is a moral cause, and should be pursued whether there are results or not, is a perennial temptation for non-profit executives—and even more for their boards. But even if the cause itself is a moral cause, the specific way it is pursued better have results. There are always so many more moral causes to be served than we have resources for that the non-profit institution has a duty—toward its donors, toward its customers, and toward its own staff—to allocate its scarce resources for results rather than to squander them on being righteous. The non-profits are human-change agents. And their results are therefore always a change in people—in their behavior, in their circumstances, in their vision, in their health, in their hopes, above all, in their competence and capacity. In the last analysis, the non-profit institution, whether it's in health care or education or community service, or a labor union, has to judge itself by its performance in creating vision, creating standards, creating values and commitment, and in creating human competence. The non-profit institution therefore needs to set specific goals in terms of its *service* to people. And it needs constantly to raise these goals—or its performance will go down.

2

Don't's and Do's— The Basic Rules

There are some don't's and some do's for non-profit institutions. Disregarding them will damage and may even impair performance.

Non-profits are prone to become inward-looking. People are so convinced that they are doing the right thing, and are so committed to their cause, that they see the institution as an end in itself. But that's a bureaucracy. Soon people in the organization no longer ask: Does it service our mission? They ask: Does it fit our rules? And that not only inhibits performance, it destroys vision and dedication.

One good example of what not to do is the way a large community hospital tackled the nursing shortage. It worked out elaborate policies to make the nurses "feel better." But the nurses' turnover only increased, and the shortage of nurses grew worse. All the measures to make the nurses "feel better" only made them more conscious of the gap between what they knew they should be doing and what the hospital allowed them to do. All the measures only made the nurses more dissatisfied.

Another hospital first asked the nurses, "How do *you* define your performance?" Every nurse said, "My contribution should be patient care." But everyone also said, "You load me down with chores and paper shuffling which have nothing to do with patient care." The solution was quite simple: Hire clerks, one for each floor, who do the chores and the paperwork. This freed the nurses for what they knew they should be doing, that is, patient care. Nurses' morale rose dramatically, turnover all but disappeared, and instead of a nurse shortage, the hospital actually found itself with a surplus. Fewer nurses carried the load and they enjoyed it. In the end the hospital could substantially raise individual nurses' pay, without running a higher nursing payroll.

In every move, in every decision, in every policy, the non-profit institution needs to start out by asking, Will this advance our capacity to carry out our mission? It should start with the end result, should focus outside-in rather than inside-out.

Dissent, as we shall see shortly, is essential for effective decision making. Feuding and bickering are not. In fact, they must not be tolerated. They destroy the spirit of an organization.

Most people think that feuding and bickering bespeak "personality conflicts." They rarely do. They usually are symptoms of the need to change the organization. It may have grown very fast and in the process outgrown its structure; nobody quite knows what he or she is responsible for. Then people begin to blame each other. I've seen this happen in an organization that was serving meals to shut-ins. That's what all the volunteers thought they were doing, and so did the people who were running the organization. But over the years, the volunteers visiting the shut-ins also took on the visiting nursing care in mobile home parks, helping lonely older people get in touch with their relatives, helping them with their Social Security, taking them to physical therapy, and so on; altogether, a dozen different kinds of help for low-income, older, handicapped people. And yet the whole organization was still based around delivering meals. Then you have constant bickering about borrowing cars from people, about being late, and about all kinds of minor things.

That's a sign that you'd better look at your organization. Are you organized for yesterday rather than today? Are you organized for the kind of small, cozy family operation you were, and now you've grown from a four-room boardinghouse into a six-hundred-room hotel without any change? When the noise level rises, it's a sign of discomfort. Your organization structure and the reality of your operation aren't congruent anymore. Then you need a change in your structure.

A final *don't:* Don't tolerate discourtesy. Since the beginning of the world, young people have resented good manners as dishonesty. They think manners are substance. If you say "Good morning" while it rains outside, you are a hypocrite. But there is a law of nature that where moving bodies are in contact with one another, there is friction. And manners are the social lubricating oil that smoothes over friction. Young people always fail to see this. The only difference is that in my youth you got slapped if you were not courteous; but we didn't feel like being courteous either. One learns to be courteous—it is needed to enable different people who don't necessarily like each other to work together. Good causes do not excuse bad manners. Bad manners rub people raw; they do leave permanent scars. And good manners make a difference.

The most important *do* is to build the organization around information and communication instead of around hierarchy. Everybody in the non-profit institution—all the way up and down—should be expected to take information responsibility. Everyone needs to learn to ask two questions: What information do I need to do *my* job—from whom, when, how? And: What information do I owe others so that they can do *their* job, in what form, and when?

When I first started working some sixty years ago, there simply was no information. Organizations had to be many-layered, tight hierarchies. Now we have enormous information capacity. This means that organizations can be much flatter and have many fewer layers. That's a great improvement. For we know that each level of management is a "relay"; and each relay in an information chain cuts the message in half and doubles the "noise." But it also means that individuals in the organization have to take information responsibility. Otherwise, we'll drown in meaningless data.

Above all, people in the information-based organization need to take responsibility for upward communication.

There is an old example. A hundred years ago, two brothers,

both surgeons in a small town in rural Minnesota, founded the first modern medical clinic—the Mayo Clinic. It was a total innovation, and everybody knew it could not work. Here were two country surgeons bringing in all kinds of high-powered specialists, and almost no layers of management. But it did work, perfectly. Every senior physician at Mayo reported directly to one of the two Doctors Mayo. And each month each Chief of Service sat down and wrote in full what was going on with each patient. In this report, he also discussed what changes were needed in the way the clinic was run or patients were treated, and where the clinic had to acquire new competence or improve its performance. And each Chief of Service, whether urologist or eye man, was expected to mobilize whatever team of physicians was needed across the whole Mayo organization to deal with whatever patient need existed. This was, of course, long before the computer.

In the information-based institution, people must take responsibility for informing their bosses and their colleagues, and, above all, for *educating* them. And then all members of the non-profit institution—paid staff and volunteers—need to take the responsibility for making themselves understood.

This requires that everyone think through and put down in writing what the organization should hold him or herself accountable for by way of contribution and results. Then, everybody has to make sure that this is understood from the bottom up, from the top down, and sideways.

This is also the one way to build mutual trust. Organizations are based on trust. Trust means that you know what to expect of people. Trust is mutual understanding. Not mutual love, not even mutual respect. Predictability. This is far more important in the non-profit organization, because typically it has to depend on the work of so many volunteers and on so many people whom it does not control.

But there are also teachers who have tenure or pastors who are nobody's "subordinates." Then you need mutual trust—and if you don't know what to expect from one another, you will soon feel let down by that fellow or that woman next door. People assume—

rightly so in a non-profit institution—that they are all dedicated to the same cause. So, when they are betrayed, or feel betrayed, it hurts much more. It's more important in the non-profit institution than it is in a business to insist on the clarity of commitments and relationships, and on the responsibility for making yourself understood and for educating your co-workers.

Everyone believes in delegation. But it needs clear rules to become productive. It requires that the delegated task be clearly defined, that there are mutually understood goals and mutually agreed-on deadlines, both for progress reports and for the accomplishment of the task. Above all, it requires clear understanding of what the person who delegates and the person who takes on the assignment expect and are committing themselves to. Delegation further requires that delegators follow up. They rarely do—they think they have delegated, and that's it. But they are still accountable for performance. And so they have to follow up, have to make sure that the task gets done and done right.

Finally, it is the duty of the person to whom a task is delegated to inform the delegator of anything unexpected that happens, and not to say, "But I can take care of it."

STANDARD SETTING, PLACEMENT, APPRAISAL

For each person to take responsibility for his or her own contribution and for being understood requires standards. Standards have to be concrete; for example, the standard for the emergency room of the hospital which I quoted earlier: everyone who comes in is seen by a qualified person in less than a minute.

Standards have to be set high; you cannot ease into a standard. When we went in to work in developing countries, we all made the same mistake. We said: Here are untrained, unskilled people, so let's start low. If you start low, you can never go higher. *Slow* is different from low. Sure, at the beginning of a new effort with a new person, you go slow. You make mistakes. But the standard is clear. There is a great deal to be said for the old schoolteacher

of mine many years ago, who put examples of beautiful penman-
ship on the wall on the first day of second grade, and said: "This
is how you are going to write." None of us kids could do it, and
most of us never did, speaking for myself. But none of us has ever
felt that sloppy handwriting is anything to be proud of.

Clear standards are particularly important in the non-profit
institution that is both centrally run and a "confederation" of
autonomous locals. Originally, there were only a few such organizations around—mostly very large ones. The oldest is, of course,
the Catholic Diocese. Then came the American Heart Association,
the Red Cross, the Scouts, and many others. Now you have hospital chains and state university systems. We have a number of large
Protestant churches which staff and support several small "outreach" churches, each with its own Vestry, its own congregation,
and its own locally raised budget. In all of them, the standards
have to be uniform across the board. But each local organization—
the council, the chapter, the parish, the diocese, the hospital—has
to be autonomous and has to make its own decisions. Squaring
these conflicting demands for autonomy and conformity requires,
above all, clear and high standards. But this kind of confederation
also requires that the central organization think through the two
or three things—not just the things to say, but the things to *do*.
In the Catholic Diocese, the bishop makes the critical personnel
decisions; he alone appoints parish priests. The Scouts provide
centrally the program material, the books for the badges, and the
innovations such as the Daisy Scouts. Headquarters also provides
the national image and handles public and governmental relations.

Next, such organizations need *control of standards*. That's the
most difficult thing to do. That's where the chief executive officer
needs not so much skill as respect, so that a local council will
accept a veto from the center even though it doesn't like it. It helps
if the central organization controls promotions in the system, the
way the bishop does in the Catholic Diocese. But in most non-
profit confederations, the local organization picks its own people.
A confederation therefore requires that the top people constantly
visit the organization's various locations—that they do so *person-*

ally rather than through staff. This is a basic requirement for the voluntary confederation, which mobilizes local energies for local performance but for a common mission that transcends local boundaries.

And the people in the central organization must remind themselves all the time: We are the servants of the local chapter, the servants of the local hospital. It is part of our job to make sure they have standards; but we are their servants. They do the work. We are not their bosses; we are their conscience.

And the people in the local chapter, the local hospital, the local parish, must remind themselves all the time: we represent the larger institution. What we do or not do, and how we do it, is seen by all our constituents as the deeds, the standards, the personality of the entire organization.

Standards should be very high and goals should be ambitious. Yet they should be attainable. Indeed, they should be attained, at least by the star performers of the institution. The non-profit institution therefore needs to work hard at placing people where they can perform. It needs to place people where their strengths are relevant to the assignment. Then, one can legitimately make demands on people.

But one also needs to use the star performers to raise the sights, the vision, the expectations, and the performance capacity of the entire organization. *One features performers.* The best way—and the way that conveys the most recognition and builds the most pride—is to use star performers as the teachers of their colleagues. Put them up front at the chapter meeting and have them tell the rest of us how they obtain their outstanding results. Nothing makes as much impact on a sales force as to have a successful salesman stand up before his peers and tell them, "This is what has worked for me." And it does even more for the star performer. There is no sweeter recognition.

People need to know how they do—and volunteers more than anyone else. For if there is no paycheck, achievement is the sole reward. Once goals and standards are clearly established, appraisal becomes possible. Sure, it's the responsibility of the superior. But with clear goals and standards, the people who do the work appraise themselves.

An appraisal should always start out with what the person has done well. Never start out with the negative: You'll get to it soon enough. But one can only base performance on strengths, on what people have got rather than on what they ain't got.

And it is the function of any organization to make human strengths effective in performance and to neutralize human weaknesses. This is its ultimate test.

THE OUTSIDE FOCUS

One more basic rule: Force your people, and especially your executives, to be on the outside often enough to *know* what the institution exists for. There are no results inside an institution. There are only costs. Yet it is easy to become absorbed in the inside and to become insulated from reality. Effective non-profits make sure that their people get out in the field and actually work there again and again.

In one of the most successful large hospitals, for instance, each staff member (including accountants and engineers) works one week a year on a floor as a nurse's aide. And each of them every other year has himself or herself actually admitted under a fictitious name and spends twenty-four hours as a patient. There is an old saying that every physician needs to have been sick and a patient to be a good doctor.

And don't let people stay forever in a staff position in the office. Rotate them regularly back into work in the field. It's an old rule of effective armies that every officer rotates back into a troop command every few years.

3

The Effective Decision

Executives, whether in a non-profit institution or in a business, actually spend little time on decision making. Far more of their time is spent in meetings, with people, or in trying to get a little information. Yet it's in the decision that everything comes together. That is the make or break point of the organization. Most of the other tasks executives do, other people could do. But only executives can make the decisions. And they either make decisions effectively or they render themselves ineffective.

The least effective decision makers are the ones who constantly make decisions. The effective ones make very few. They concentrate on the important decisions. And even people who work hard on making decisions often misapply their time. They slight the important decisions and spend excessive time making easy—or irrelevant—decisions.

The most important part of the effective decision is to ask: What is the decision really about? Very rarely is a decision about what it seems to be about. That's usually a symptom.

Some twenty years ago, a Girl Scout Council in a major suburban area realized that the ethnic composition of the area was changing rapidly. It had been lily-white, and so had the Scouts. But now the area was rapidly becoming highly diverse: blacks, Hispanics, Asians were arriving in large numbers. That the Council had to offer scouting to the children of the newcomers was obvious to everyone. But so was the enormous cost of providing scouting to very poor neighborhoods. The question that seemed to demand a decision was, therefore, seen as a financial one: How do we raise the money? And the answer to that question seemed obvious: Have separate troops for different ethnic groups. Otherwise, it was feared, financial support from the affluent group, the whites, might be endangered.

Fortunately, one of the leaders then asked: What is this decision all about? Is our mission to raise money, or is it to build a nation? It was clear at once that the decision was one of basic principle, to be decided contrary to all of the Council's precedents. The answer had to be that, whatever the financial risk, we are not going to have ethnic troops. That is the past. We have to emphasize that young women are young women—not black, not white, not Italian, not Jewish, not Vietnamese—but young American women. That is what the decision was really all about. Once this was clear, the decision made itself. And the whole community accepted that decision without a murmur, once it was explained.

A major university with severe budget problems had to accept that it must cut programs. But which ones? At first, this was seen as a financial decision: where do we spend the most? The ensuing civil war within the faculty almost destroyed the institution. But then one board member said, "We are tackling the wrong issue. We should be discussing whether to put our major emphasis on the continuing education of adults or whether to stick with teaching the young. That's what this decision is about. The rest is implementation." Suddenly it became clear why people had been so hot under the collar. The decision was not about the budget but about the future of American higher education and the university's role in it, and this is something on which good people *should* disagree. Such a decision is a strategic decision, and halfway measures won't do. If the university's future is in continuing education, it is not going to cut. It has to go out and raise the money; otherwise, it has no future.

Decisions always involve risk taking. And effective decisions take a lot of time and thought. For this reason, one doesn't make unnecessary decisions. Again and again, non-profit institutions go through a painful reorganization, moving staff and activities around because two people are feuding with one another. But they have been feuding for twenty years and will keep on feuding whatever the organization structure. Leave them alone.

And don't make decisions on trivia. I live sixty miles east of Los Angeles, with four freeways into the city. They all have the same mileage; it's totally unpredictable which one will be jammed. Whether you take one or the other is not a decision. Routine decisions are decisions that have no consequences, or at least no foreseeable consequences. Don't waste time on them.

OPPORTUNITY AND RISK

The next question in decision making is opportunity versus risk. One starts out with the opportunity, not with the risk: If this works, what will it do for us? Then look at the risks. And there are three kinds of risks:

There is the risk we can afford to take. If it goes wrong, it is easily reversible with minor damage. Then there is the irreversible decision, when failure may do serious harm. Finally there is the decision where the risk is great but one cannot afford *not* to take it. Here's an example. Forty years ago a Brooklyn neighborhood in New York radically changed from white working class to a black slum. A major hospital in the area almost overnight became empty, going down to about 12 percent occupancy. Its regular physicians had left with their patients. Keeping the hospital open could not be economically justified but the community needed its services. The decision—and it was bitterly fought—was to keep the hospital open and to raise the money somehow for the three to five years until the hospital's patient base could be rebuilt. The decision came very close to total disaster. But to stay open was a risk the hospital had to take if it wanted to maintain its mission.

THE NEED FOR DISSENT

All the first-rate decision makers I've observed, beginning with Franklin D. Roosevelt, had a very simple rule: If you have consensus on an important matter, don't make the decision. Adjourn it so that everybody has a little time to think. Important decisions are risky. They *should* be controversial. Acclamation means that nobody has done the homework.

Because it is essential in an effective discussion to understand what it is really about, there has to be dissent and disagreement. If you make a decision by acclamation, it is almost bound to be made on the apparent symptoms rather than on the real issue. You need dissent; but you have to make it productive.

About seventy years ago, an American political scientist, Mary Parker Follet, said that when you have dissent in an organization, you should never ask *who* is right. You should not even ask *what* is right. You must assume that each faction gives the right answer, but to a different question. Each sees a different reality.

A few years ago, as we saw earlier, a major hospital was torn by internal conflict within its medical staff. One group advocated moving the eye clinic out of the hospital. Most eye operations have become ambulatory and it is far more economical to do them where they do not have to carry the whole overhead of the big hospital. The other group saw such a move as the first step toward complete restructuring of the hospital. Both were right, but both saw only part of the reality.

Instead of arguing what is right, assume that each faction has the right answer. But which question is each trying to answer? Then, you gain understanding. You also gain, in many cases, the ability to bring the two together in a synthesis. Then you can say: In this case we are not deciding on ophthalmology; that is just an incident. But the decision to move the eye clinic out commits us to restructuring the hospital. If we believe that moving out of the hospital is tomorrow's right structure, let's not talk economics, whether of the hospital or of eye surgery. And everybody understands it. Look upon dissent as a means of creating understanding and mutual respect.

Emotions always run high over any decision in which the organization is at risk if that decision fails, or in one that is not easily reversible. The smart thing is to treat this as constructive dissent and as a key to mutual understanding.

If you can bring dissent and disagreement to a common understanding of what the discussion is all about, you create unity and commitment. There is a very old saying—it goes back all the way to Aristotle and later on became an axiom of the early Christian Church: In essentials unity, in action freedom, and in all things trust. And trust requires that dissent come out into the open, and that it be seen as honest disagreement.

This is particularly important for non-profit institutions, which have a greater propensity for internal conflict than businesses precisely because everybody is committed to a good cause. Disagreement isn't just a matter of your opinion versus mine, it is your good faith versus mine. Non-profit institutions, therefore, have to be particularly careful not to become riddled by feuds and distrust. Disagreements must be brought out into the open and taken seriously.

A second reason to encourage dissent is that any organization needs a nonconformist. If and when things change, it needs somebody who is willing and able to change. This is not the kind of person who says, "There is a right way and a wrong way—and our way." Rather, he or she asks, "What is the right way *now?*" You don't want only yes-men or yes-women. You want a critic—and one the organization respects.

Bringing disagreement into the open also enables non-profit executives to brush aside the unnecessary, the meaningless, the trivial conflict. It enables them to concentrate on the real issues. When you bring conflicts out in the open, a good many disappear. People realize that they are trivia and not that serious. Yes, there *is* a conflict. You here in the surgery see one thing and you here in internal medicine see another. But is this pertinent to *this* specific case? If not, you say what our teacher of religion said to us when we were thirteen: "Boys, kill each other, but not in my class." Fight it out outside; it doesn't belong here. You don't resolve the conflict, but you do make it irrelevant. If you can do that, you are way ahead.

Another example: I was present, not so long ago, at a meeting at a museum that degenerated into civil war. People were screaming at each other until one of the wise old men pointed out that the two groups were both right. One, in arguing for a big new building, assumed the kind of museum we are now building, which is a museum that is a community asset. So, members of this group assumed we were talking about a tremendous expansion. The other group assumed the opposite. It wanted to concentrate on a very small number of real masterpieces and create a standard of excellence in which every single object was the best in its class, which is very much the way the great nineteenth-century collectors went about their business. The word "museum" was the same, but that was the only thing.

Once the position of each group was understood, it became clear that the conflict had nothing to do with the matter under discussion. Sooner or later a decision will be made to go one way or the other, and then half the board will resign—maybe to start a new museum. But that wasn't what we had to decide at that meeting. Suddenly there was peace, harmony, even laughter.

CONFLICT RESOLUTION

You use dissent and disagreement to resolve conflict. If you ask for disagreement openly, it gives people the feeling that they have been heard. But you also know where the objectors are and what their objections are. And in many cases you can accommodate them, so that they can accept the decision gracefully. That also enables them very often to understand the arguments of the winning side. Maybe not to *accept* them; but to see that these people are neither stupid nor malicious. They only differ. In this way you resolve conflict. You do not prevent disagreement, but you do resolve conflict.

Another way to resolve conflict is to ask the two people who most vocally oppose each other, especially if both of them are respected community members, to sit down and work out a common approach. They do this by starting out with the areas in which they agree.

The third way is by defusing the argument. You say, "Let's start out by finding out what we *agree* on." Then disagreements often turn out to be peripheral. On essentials there is common ground and you can work out things. In some cases you say, "Let's split the difference," or, "Let's postpone this," or, "Is this really that important?" You play down the areas of disagreement and play up the areas of agreement.

These are by no means new techniques; there are examples in the Old Testament. Finding common ground especially is what the elders of any tribe do to maintain unity. One cannot prevent conflict. But one can make it—I wouldn't say irrelevant, but secondary. And the best tool for this is the constructive use of dissent.

FROM DECISION TO ACTION

A decision is a commitment to action. But far too many decisions remain pious intentions. There are four common causes for this. One is that we try to "sell" the decision rather than to "market" it. In the West, we tend to make the decision fast—and then we start to "sell" it to the people in the organization. That takes three years, and by the time the decision has been "bought," it has become obsolete. Here we can learn from the Japanese. They build the implementation in *before* they make the decision. In the Japanese organization, everyone who will be affected by the decision—and especially everyone who will have to do something to carry it out—is asked to comment on the issue before that decision is made. This looks incredibly slow. Westerners watching the process climb up the walls. But then the Japanese make the decision—

the point at which we in the West begin to "sell." Not so the Japanese. Bingo! The next day everyone understands it, everyone acts on it.

A second way to lose the decision is to go systemwide immediately with the new policy or the new service. This jumps the testing stage. We disregard what Frances Hesselbein of the Girl Scouts told us in her interview in Part One of this book: Find the targets of opportunity in your non-profit institution and concentrate on them. Don't try to convert everybody right away.

I like to try the new in three different places with three different people—something I learned forty years ago from the people who introduced physical therapy in the American hospital. There was almost universal resistance to the idea. Most hospitals said it was none of their business. The innovators didn't even try to convert the non-believers. They picked three hospitals in three communities that were eager to do physical therapy: a large teaching hospital with many older people, stroke victims, and so on; a small semi-rural hospital that had lots of industrial and farming accidents; and a fair-sized suburban community hospital with a lot of ordinary cases, broken bones, arthritis, and so on. They worked only with these three hospitals for five years. By then, every hospital in the country wanted physical therapy.

But by then, also, the product had become quite different from the original design. The three pilots showed, for instance, that psychological counseling and work with the patient's family are just as important in rehabilitation as exercise and physiology—something which had not even occurred to the innovators but which made an enormous difference in effectiveness. In industry we learned long ago that we are going to be in trouble if we jump the pilot stage. We have to learn that this is just as true for social projects and services.

The third caveat: no decision has been made until someone is designated to carry it out. Someone has to be accountable—with a work plan, a goal, and a deadline. Decisions don't make themselves effective; people do.

Finally—common mistake number four—I've seen wonderful

decisions come a cropper because nobody really thought through who had to do what. In what form should the decision be communicated to each person who has to implement that decision so that he or she can actually act? What training does each need? What tools? I have seen a decision couched in a brilliant mathematical model which forklift drivers in the warehouse were expected to carry out. It didn't become effective. You not only have to translate a decision into the language of the people who have to do the work; you also have to fit it into their assumptions. You have to build the new behavior into their instructions, their training, their compensation. And then you have to follow up. Don't depend on reports. Go to the warehouse and look. Otherwise, you'll find a year later that nothing has happened.

Every decision is a commitment of present resources to the uncertainties of the future. This, according to elementary probability mathematics, means that decisions will turn out to be wrong more often than right. At the least they will have to be adjusted. Practically every single decision American hospitals made in the sixties and seventies has been shot out of the water by changes in government—particularly reimbursement policies on Medicare. As a result, hospitals suddenly have a surplus of beds. But that's a normal outcome for decisions on the future.

The decision always has to be bailed out. That requires two things. First, that you think through alternatives ahead of time so that you have something to fall back on if and when things go wrong. Second, that you build into the decision the responsibility for bailing it out, instead of going in and arguing about who made what mistakes. One weakness of non-profit institutions is that they believe that they have to be infallible—far more so than businesses. Businesses somehow know mistakes are being made. In non-profit institutions, mistakes are not permitted. And so if something goes wrong, a court-martial begins. Whose fault is it? Instead, we need to ask, Who is going to bail this out? Who is going to redirect the program or operation, and how?

4

How to Make
the Schools Accountable

Interview with Albert Shanker *

PETER DRUCKER: Albert, you have been leading a crusade to improve performance in the classroom, to make teachers and schools accountable for performance, and to build the school around the classroom teacher.

How do you define performance in the school?

ALBERT SHANKER: The way to deal with this is to ask: What kind of human being are we trying to produce? Most educators deal with the question very narrowly in terms of test scores, SAT scores, or narrow performance. But essentially performance in education occurs along three dimensions. One, of course, is knowledge. The second dimension, I would say, is being able to enter the world as a participating citizen and perform within the economy. The third has to do with the growth of the individual and participation in the cultural life of society.

Unfortunately, we don't do a very good job of even getting close to measuring these gains.

PETER DRUCKER: But it makes sense to say that unless a person has those tangible, measurable, knowledge skills, a foundation is

*Albert Shanker is president of the American Federation of Teachers AFL–CIO.

131

lacking. Somehow, one has to set priorities for defining what achievement is.

ALBERT SHANKER: I think the priority is to assess achievement longer range. When you measure small gains each semester or each year, you get down to things that don't mean very much. Rather trivial things that a student can study for an exam. They don't mean anything a week later. They're not even remembered later on.

PETER DRUCKER: I think I'm a living example of this. My school grades were always excellent. I learned very little and studied less, but I knew how to take exams.

ALBERT SHANKER: Let me illustrate what learning is not and what it is. Teachers are required to give a course in Nature, so they put bird charts around the room. They show flash cards and have the children give the names of the birds. The end result is an examination where the students regurgitate the names of the birds. But the kids don't remember the names very long; all that's there a few months later is a permanent dislike of birds.

In the Boy Scouts, when I was a youngster, they had a bird-study merit badge. You actually had to *see* forty different birds. You soon find you can't do that by walking across the street to a park. You have to get up early in the morning and go to a swamp or woods. You don't want to do it alone, so you find one or two friends who will go with you. Soon you find that the birds you see out there don't look the way they do in pictures. What happens over the months of going out with your friends and looking at these birds is you begin to feel a sense of power. You can see birds around you that no one else can see.

A key problem for schools is to organize learning for youngsters in such a way that it doesn't become something memorized and instantly forgotten, but something that becomes part of you. I have never met anyone who went through this experience in the Boy Scouts for whom it didn't remain a pretty lasting interest.

PETER DRUCKER: The implication of this is, first, that you put the learning responsibility on the student rather than the teaching responsibility on the teacher. Is that central to the way you see performance?

ALBERT SHANKER: Essentially, the way schools are organized is to get a lot of activity and work on the part of teachers while the students sit and, you hope, listen. You hope that they are remembering something. And you create a few punishments or rewards in terms of grades or leaving students back. Without that responsibility and without that engagement by students, the results are very, very meager.

PETER DRUCKER: For hundreds of years, then, our emphasis has been on how well the teachers teach rather than on how well the student learns?

ALBERT SHANKER: The school is organized on the assumption that the student is a thing to be worked on, not that the student is the worker. A school is something like an office. That is, the students are required to read reports and write reports. It's more like an office than any other place. But it's an office in which the student is given a desk and told, "Your boss there, the teacher, will tell you what to do. But every forty minutes you will move to a different room and you will be given a different desk and you will be given a different boss who will give you different work to do." Now, no one would organize an office that way. The student is not being viewed as a worker who has to be engaged, but as raw material passing through a factory. Well, of course, it doesn't work because that's not the way the process of learning goes on.

PETER DRUCKER: I've been a teacher-watcher since fourth grade, when I had the great good luck of two exceptional teachers. And I've been a teacher myself since I was twenty. I have yet to see a great teacher who teaches *children*. All the great teachers I've seen made no distinction between children and adults. Only the speed is different. Whatever the task is, you do it on an adult level. The task may be a beginner's task; the standards are not. The fourth-grade teacher whom I still remember once said many years later that there are no poor students; there are only poor teachers. That would imply that the job of the teacher is to find the strengths of the student and put them to work, rather than to look at the student as somebody whose deficiencies have to be repaired.

ALBERT SHANKER: When I taught, I was very rarely approached by a principal or assistant principal and asked whether the children were really learning or really engaged. I had a very tough class, mostly youngsters that had just flown in from Puerto Rico, who had great difficulties with the language. I was hoping that someone would come in to help me. Then, the door opened one day and there was the principal. After what seemed to me like a half hour, but must have been maybe thirty seconds, he said: "Mr. Shanker, there are a lot of pieces of paper on the floor throughout your room. That's very unprofessional. Would you see to it that they're picked up?" Then the door closed and he went away. The only thing that anyone was ever interested in was essentially a set of bureaucratic requirements.

PETER DRUCKER: One implication is that the school has to be focused on performance and results rather than on rules and regulations and, therefore, needs a clear definition of its mission.

ALBERT SHANKER: It needs that. And it also needs a system to accomplish that. One can't expect school board members *not* to be responsive to their constituents. One can't expect a school superintendent *not* to be concerned with how he looks in front of the public and whether his contract gets renewed.

PETER DRUCKER: Now if I may move to your own work in your own organization, that big union you have built. When you took national office sixteen years ago in 1974 as chief executive officer of what was then a fast-growing and very controversial union, which had a very difficult time in the 1960s, what was the first thing you did?

ALBERT SHANKER: The first thing I did was to try to move the union away from its orientation during the previous fifteen years. Let me take one step back. When I started to build the union as a teacher, and later as a staff member, the toughest thing I had to do was convince teachers that they had a right to pursue their own self-interest economically. The notion of belonging to a union as against a professional association was just anathema. However, by the time I became president of the American Federation of Teachers, it had gone too far in that direction. Teachers were viewed as people who went on strike every year—not interested in the children, not interested in educational issues. There was a tremendous backlash. As a result of the GI Bill and the expansion of higher education in the United States, we also had a much more educated public that was far more critical of the public schools. The image of schools and teachers had gone down and we faced threats of privatization, threats of tuition tax credits, of vouchers, of the public finding alternatives to public education.

The first thing I worked for at that time was to develop new alliances with the business community. We had to have a magazine that was a professional journal, not a union journal. We had to be viewed not just as people who have the guts to fight and to strike, but as people who are teachers and who have knowledge, because otherwise our entire industry will go down.

Our industry going down has a much broader impact than will it hurt the union, or will it hurt school boards. Public education in this country is the place where people of different races and religions come together. It's what we used to call the institution that "Americanizes," a rather old-fashioned word. In this country if this institution goes down, it's not just a narrow problem for the American Federation of Teachers. It's a broad problem, because our private schools tend to be Catholic, Protestant, Jewish, Black, Hispanic, language-oriented, even politically oriented. What would be the consequences for the future of the country if the overwhelming majority of children in the future were brought up only with their own kind? So, our orientation had to move away from confrontation and, in a sense, toward saving the institution, which I saw—and which I still see—as one in great danger.

PETER DRUCKER: You know, Albert, you have talked about one of the key problems in running any organization—balancing long-run and short-run objectives. When you moved in, you had to introduce a long-run objective in which the survival and success of the institution becomes the long-run critical point. On the other hand, you had to maintain the intermediate goal of defending the teachers' immediate interest in next year's contract. How do you balance those two?

ALBERT SHANKER: It's very tough. We know that teachers need a union if they're going to engage in conflict. But do they need a union to cooperate with management? We don't know yet.

PETER DRUCKER: What you said just now is important. It's important for the whole union movement and not just in this country. In every developed country, the labor union is faced with that problem. But it isn't just an issue for unions. International charitable organizations get an immediate outpouring of funds by showing starving children in Ethiopia. But it is terribly hard to get support to *prevent* the Ethiopian famine and to do development work, where results take eight or ten years. That problem is likely to create a tendency in the staff to say, "Don't talk about long-range goals; it only confuses people. Let's play on the heart strings and show starving babies."

That's self-defeating in the end. After five or eight years, people get awfully tired of it. I've been working with hospitals where we have been saying for twenty years the long-run goal is to get patients out of the hospital, not in. If we don't do it, the way medicine is going, we'll be in a severe crisis. Everybody said, yes, that's the long-range goal, but don't let's talk about it. Doctors don't want to hear it; nurses don't want to hear it; the donors don't want to hear it. Most hospitals are in desperate straits because they were totally unprepared when the patients began to be treated outside the hospital. But the few hospitals that actively worked on creating outreach clinics are doing well.

ALBERT SHANKER: That's exactly the experience we're beginning to have with some schools. Those who are pursuing the long-term rather than the short-term objective find that the short-term objective falls into place. In Rochester, New York, for instance, union and management stuck their necks out several years ago and decided to put in some very controversial programs. They included experienced teachers training new teachers; peer review; deciding which teachers would train other teachers and evaluate them, and ultimately decide that some of them couldn't make it at the end of their probationary period. We tried the same kind of program in Toledo, Ohio. These are both districts that had a lot of conflict. They'd had strikes; they had people starting to move out of the school district or into private schools. And the radical turnaround in the relationship between the union and management and what they were willing to do to change the roles and relationships of people shocked the public into awareness. People in the business community said, "We ought to support this." Newspapers started to support it.

The result is that in each of these cities, the city governments and the local unions came to agreement on spectacular contracts in terms of salary. The recent Rochester contract provides that in three years the top teachers will earn close to $70,000 a year. In the previous contract, the top was about $40,000 a year. That is now providing a spur to others. This, now, is the way to start doing things very, very differently and to show a basic commitment to the enterprise.

PETER DRUCKER: Basically, the implication of this experience for non-profit institutions is to keep an eye on the fundamental, long-term goal. Make sure you move toward it, and you'll gain credibility. And be sure you define performance and hold yourself accountable for it.

ALBERT SHANKER: That's right. I think the public may have given up on many of our public institutions because of a feeling that these people have their jobs, their security, their tenure, their Civil Service regulations; but they've really stopped trying. They're just doing what they did last week and last year and five years ago, whether it works or not.

PETER DRUCKER: And in many cases, alas, they are right.

ALBERT SHANKER: That's correct. They are right. But even an old institution like the school can be turned around.

5

Summary:
The Action Implications

Performance is the ultimate test of any institution. Every non-profit institution exists for the sake of performance in changing people and society. Yet, performance is also one of the truly difficult areas for the executive in the non-profit institution.

I'm always being asked what the differences are between business and non-profit institutions. There are few, but they are important. Perhaps the most important is in the performance area. Businesses usually define performance too narrowly—as the financial bottom line. If that's all you have as a performance measurement and performance goal in the business, you are not likely to do well or survive very long. It's too narrow. But it's very specific and concrete. You don't have to argue about whether we are doing better because results within terms of profitability or market standing or innovation or cash flow are easily quantifiable and very hard to ignore.

In a non-profit organization, there is no such bottom line. But there is also a temptation to downplay results. There is the temptation to say: We are serving in a good cause. We are doing the Lord's work. Or we are doing something to make life a little better for people and that's a result in itself. *That is not enough.* If a business wastes its resources on non-results, by and large it loses its own money. In a non-profit institution, though, it's somebody else's money—the donors' money. Service organizations are accountable to donors, accountable for putting the money where the results are, and for performance. So, this is an area that needs special emphasis for non-profit executives. Good intentions only pave the way to Hell.

Nonetheless, non-profit institutions find it very hard to answer the question: What, then, are "results" in our institution? It can be done, however. Indeed, results can even be quantified—at least some of them. The Salvation Army is fundamentally a religious organization. Nevertheless, it knows the percentage of alcoholics it restores to mental and physical health and the percentage of criminals it rehabilitates. It is highly quantitative. For many organizations in the non-profit sector, to be specific about results is still odious. They still believe their work can only be judged by quality—if at all. Some of them still quite openly sneer at any attempt to ask: "How well are you doing in terms of the resources you spent? What return do you get?" One sometimes has to remind them of the Parable of the Talents in the New Testament: Our job is to invest the resources we have—people and money—where the returns are manifold. And that's a quantitative term.

There are different kinds of results. First, you have immediate results. Then, you have the long-term job of building on those first results. Maybe it's not easy to define precisely what results you have, but it's got to be done in such a way that one can ask: "Are we getting better? Are we improving?" And: "Do we put our resources where the results are?"

We need to remind ourselves again and again that the results of a non-profit institution are always outside the organization, *not* inside. Results for the Salvation Army are among the alcoholics and the prostitutes and the hungry. Results for the schoolteacher are kids who learn.

And can good intentions and hopes ever justify non-results? A few Jesuit Fathers managed to sneak into China as missionaries in the seventeenth and early eighteenth centuries. They were brilliant men; they endured persecution and hardships and dangers. They worked terribly hard and they stayed in China year after year after year—with no results. Yet they kept on hoping, kept on trying to find a few people who would be receptive to Christianity. In the process they became very respected men in China—astronomers, mathematicians, painters. But it was a misallocation of very scarce resources to work that produced no results. In Heaven there is joy over one sinner who repents. But in Heaven, there is also, I am sure, joy over the right allocation of resources to the mission, to the goals, to results. And the Jesuits long ago stopped wasting brilliant members of their order on hopes.

One starts with the mission, and that is exceedingly important. What do you want to be remembered for as an organization—but also as an individual? The mission is something that transcends today, but guides today, informs today. The moment we lose sight of the mission, we begin to stray, we waste resources. From the mission, one goes to very concrete goals.

Only when a non-profit's key performance areas are defined can it really set goals. Only then can the non-profit ask: "Are we doing what we are supposed to be doing? Is it still the right activity? Does it still serve a need?" And, above all, "Do we still produce results that are sufficiently outstanding, sufficiently different for us to justify putting our talents to use in that area?" Then, you can do the *next* important thing, which is every so often to ask: "Are we still in the right areas? Should we change? Should we abandon?" The Salvation Army began, 128 years ago, by building shelters for the streetwalkers of London. Nobody cared then about those unfortunate women, any number of whom were poor country girls adrift in the big city. The Salvation Army still has a program to look after prostitutes. But it has given up providing hostels to shelter innocent and ignorant country girls. Those country girls now come equipped with employable skills, and they are by no means ignorant anymore; they are just as sophisticated as anybody else. So, the Salvation Army abandoned this mission even though it was the original activity.

One needs to define performance for each of the non-profit's key areas. Think through the key performance areas for this organization—not for *an* organization—for *this* one, and focus on each of them.

In a non-profit institution, where people want to serve a cause, you always have the challenge which Max De Pree discussed in his earlier interview: getting people to perform so that they grow on their own terms. They are then accomplished and fulfilled, and that makes its way down to the performance of the organization. This is essential.

Results are achieved, too, by concentration, not by splintering. That enormous organization the Salvation Army concentrates on only four or five programs. Its executives have the courage to say, "This is not for us. Other people do it better." Or, "This is not really what we are good at." Or, "This is not where we can make the greatest contribution. It does not really fit the strength we have." One of the most important things for a non-profit executive to be able to acknowledge is that "there we are not competent; we can only do harm. Need alone does not justify our moving in. We must match our strength, our mission, our concentration, our value."

Good intentions, good policies, good decisions must turn into effective actions. The statement, "This is what we are here for," must eventually become the statement, "This is how we do it. This is the time span in which we do it. This is who is accountable. This is, in other words, the work for which we are responsible." Effective organizations take it for granted that work isn't being done by having a lovely plan. Work isn't being done by a magnificent statement of policy. Work is only done when it's done. Done by people. By people with a deadline. By people who are trained. By people who are monitored and evaluated. By people who hold themselves responsible for results.

The ultimate question, which I think people in the non-profit organization should ask again and again and again, both of themselves and of the institution, is: "What should I hold myself accountable for by way of contribution and results? What should this institution hold itself accountable for by way of contribution and results? What should both this institution and I be remembered for?"

PART FOUR

People and Relationships

*your staff, your board,
your volunteers, your community*

1

People Decisions

People decisions are the ultimate—perhaps the only—control of an organization. People determine the performance capacity of an organization. No organization can do better than the people it has. It can't reasonably hope to recruit and hold much better people than anybody else, unless it is a very small organization, let's say a string quartet. Otherwise it can only hope to attract and hold the common run of humanity. But an effective non-profit manager *must* try to get more out of the people he or she has. The yield from the human resource really determines the organization's performance. And that's decided by the basic people decisions: whom we hire and whom we fire; where we place people, and whom we promote.

The quality of these human decisions largely determines whether the organization is being run seriously, whether its mission, its values, and its objectives are real and meaningful to people rather than just public relations and rhetoric.

The rules for making good people decisions are well established, though, alas, very few of us follow them correctly. Any executive who starts out by believing that he or she is a good judge of people is going to end up making the worst decisions. To be a judge of people is not a power given to mere mortals. Those who have a batting average of almost 1.000 in such decisions start out with a very simple premise: that they are *not* judges of people. They start out with a commitment to a diagnostic process.

Medical educators say their greatest problem is the brilliant young physician who has a good eye. He has to learn not to depend on that alone but to go through the patient process of making a diagnosis; otherwise, he kills people. An executive, too, has to learn not to depend on insight and knowledge of people but on a mundane, boring, and conscientious step-by-step process.

Properly done, the selection process starts with an *assignment*—not merely with a job description but an assignment. Next, the executive forces himself or herself to look at more than one person. All of us think we know who the "right" person is, as a rule. But effective non-profit executives shouldn't decide impulsively. They should look at several people so they have a safeguard against being blinded by friendship, by prejudice, or merely by habit. Thirdly, while reviewing candidates, the focus must always be on performance. Don't start with personality. Don't start with the usual silly questions such as does he get along with people, or does she have initiative? These characteristics may be meaningful in describing a personality, but they don't tell you how people perform. The right questions are: How have these people done in their last three assignments? Have they come through? Then, fourth, look at people's specific strengths. What have they shown they *can do* in their last three assignments?

Once you come to the conclusion, yes, Mary Ann is the right person, go—the final step—to two or three people with whom she has worked. If they all say, My only regret is that Mary Ann no longer works for me, then go ahead and make the job offer. But if they say, I wouldn't take her back, start thinking again.

Selecting a person to carry out an assignment does *not* end the decision process. The second stage comes ninety days later, when you call that newly appointed person in and say: Mary Ann, you have now been on this new job ninety days. Think through what you have to do to be successful, and come back and tell me. When she returns with her report, you can finally judge whether you have selected the right person for the assignment.

HOW TO DEVELOP PEOPLE

Any organization develops people; it has no choice. It either helps them grow or it stunts them. It either forms them or it deforms them. Fortunately for us as a nation, even though formal schooling in the United States has gone downhill over the last forty years, informal learning and training have exploded. These activities are now as big, in terms of both people enrolled and money spent, as formal schooling. In fact, I wish we could translate into the schools some of the lessons learned by large non-profit institutions in training people. The best of these have learned how to appraise and judge performance, and then use these tools to make each job bigger, to scale up demands, and to innovate.

What do we know about developing people? Quite a bit. We certainly know what *not* to do, and those don't's are easier to spell out than the do's. So, don't make the obvious mistakes. First, one doesn't try to build on people's weaknesses. Schools, of necessity, focus on what the kid can't do. When you're called in to a conference with the teacher of your fourth-grade child, the teacher is unlikely to say, "Your Johnny writes very well; he ought to do more writing." She's more likely to say, "Your Johnny is weak in arithmetic; he needs work on the multiplication table." That's okay from the point of view of the school because the school doesn't know what that child is going to do ten, twenty, or thirty years later. So it has to give him or her the basic skills and work on the weaknesses. But if you want people to perform in an organization, you have to use their strengths—not emphasize their weaknesses. By the time people come to work, their personalities are set. One can expect adults to develop manners and behavior and to learn skills and knowledge. But one has to use people's personalities the way they are, not the way we would like them to be.

A second don't is to take a narrow and shortsighted view of the development of people. One has to learn specific skills for a specific job. But development is more than that: it has to be for a career and for a life. The specific job must fit into this longer-term goal.

Another thing we now know is not to establish crown princes. It used to be very fashionable (and still is today in some organizations) to evaluate the new young hires and pick out the "comers." I have been working with organizations now for around fifty years and my experience is that the correlation between the high-promise people at age twenty-three and the performers at age forty-five is very poor. Lots of people I know who are world beaters at age fifty were drab and dull when they were twenty-three. Lots of high flyers come out of business schools at the top of their class and are burnt out six years later. Look always at performance, not at promise.

One of the most successful developers of people I know is the pastor of a large church. An amazing number of first-rate leaders have come out of his church, so I once asked him to explain how his church has become the breeding ground, the cradle of volunteer leaders. He told me the church tries to provide four things to young people who show up for services: (1) a mentor to guide him or her; (2) a teacher to develop skills; (3) a judge to evaluate progress; and finally, (4) an encourager to cheer them on. I then asked him which of those four roles he took for himself, and he answered: "I am the encourager. Nobody else can really do that except the person at the very top. It's an urgently needed source of help to these young people because I want people to make mistakes. They can't develop otherwise. So when they fall flat on their faces, somebody has to pick them up and say, go on. That's *my* role."

With the focus on performance rather than potential, the non-profit executive can make high demands. One can always relax standards, but one can never raise them. So, with the beginner, take more time. Make things easy. He may have to try again and again, but there is only one standard for quality performance and he has to meet it.

There are two rules I've learned that help me understand what needs to be done. One is the slogan of the Association of the Handicapped: "Don't hire a person for what they can't do, hire them for what they *can* do." You put blind people where you need sensitivity to voice, where blindness is a tremendous asset. The other piece of wisdom I learned when I was eleven. My piano teacher, in utter exasperation, said to me, "Look, Peter, you'll never play Mozart the way the great pianists play, but there is no reason why you can't do your scales as well as they do."

Next, the non-profit executive must learn how to *place* people's strengths. A very great leader of men, General George C. Marshall, Chief of Staff of the U.S. Army during World War II, had the most remarkable record in putting people into the right place at the right time. He appointed something like six hundred people to positions as General Officer, Division Commander, and so on, almost without a dud. And not one of these people had ever commanded troops before. A discussion would come up, and Marshall's aides would say, "Colonel So-and-So is the best trainer of people we have, but he's never gotten along with his boss. If he has to testify before Congress, he'll be a disaster. He's so rude." Marshall would then ask, "What is the assignment? To train a division? If he's first rate as a trainer, put him in. The rest is my job." As a result, he created the largest army the world had seen, 13 million people, in the shortest possible time, with very few mistakes.

The lesson is to focus on strengths. Then make really stringent demands, and take the time and trouble (it's hard work) to review performance. Sit down with people and say: This is what you and I committed ourselves to a year ago. How have you done? What have you done well?

For all this to come together, the mission has to be clear and simple. It has to be bigger than any one person's capacity. It has to lift up people's vision. It has to be something that makes each person feel that he or she can make a difference—that each one can say, I have not lived in vain.

The worst thing an organization can do is limit its development of people by importing society's class system into its own operations, like organizations today that decide very early which are the comers, or that you are not going to get any place if you don't have an MBA from the Harvard Business School. Performance is what counts. Not in one job, but in a series of jobs, because people are not that predictable. You may put somebody into a specific job and the chemistry is wrong, it doesn't work. People don't always get along with a boss. So, you try them in another job. The old rule is, if they try, work with them. If they don't try, you're better off if they work for the competition.

One of the great strengths of a non-profit organization is that people don't work for a living, they work for a cause (not everybody, but a good many). That also creates a tremendous responsibility for the institution, to keep the flame alive, not to allow work to become just a "job."

Hospitals, it seems to me, do the poorest job of keeping that spirit alive. So many jobs there are just routine. Partly because people do need protection against the suffering, they become callous. The leadership challenge in a hospital—for a good administrator, for a good director of nursing—is to bring people from half a dozen departments together again and again and ask: What can we be proud of? Have we really made a difference? We've had six cardiac arrests in one night and not one of the patients died. Focus on success.

There's a children's cancer ward not very far from where I live, in the City of Hope in Pasadena, where the atmosphere is joyous because the focus is on the success: on making dying, suffering children enjoy their childhood. Everybody feels the mission, despite the hard grind. Much of the work is just wiping up after vomiting children. And yet there is a sense of doing something important.

That sense of mission should be a tremendous source of strength for any non-profit organization. But it comes with a price tag. The non-profit executive is always inclined to be reluctant to let a non-producer go. You feel he or she is a comrade-in-arms and make all kinds of excuses. So, let me repeat the simple rule once more: If they try, they deserve another chance. If they don't try, make *sure* they leave.

Effective non-profit organizations also have to ask themselves all the time: Do our volunteers grow? Do they acquire a bigger view of their mission and greater skill? They look at the people who work for them not as a static resource, but as a dynamic, growing force. In many ways, the successful institutions do as the Girl Scouts does. They measure themselves as much by the development of their staff and volunteers as by the development of the young girls. Make sure that volunteers are given responsibility; they must be able to spread their wings and have autonomous commands. In the Scouts, they start as troop leaders and camp leaders, as teachers for badges. Then they receive task force assignments, are asked to lead teams and to develop materials. Next, they move into leadership positions with the local and the national organization.

The most important way to develop people is to use them as teachers. Nobody learns as much as a good teacher. Selecting someone to be a teacher is also the most effective recognition. Whether you talk to salesmen or Red Cross workers, you'll find that no recognition is sweeter than to be asked: "Tell us what you do to be so effective?"

The final development tool is needed less for volunteers than it is for regularly employed staff workers, who can so easily become inbred and ingrown. Push them outside: for example, into adult education at the local high school or college.

It is a common complaint that many bosses do not really want top-performing subordinates because they put pressure on them. That's just what an effective organization *does* want, and that's where a volunteer organization has an advantage. The volunteer who performs isn't out to get the paid executive's job, as a rule, and is not seen as a threat. There's an old story about the symphony orchestra which a great composer, Gustav Mahler, built in Vienna a hundred years ago, just before the turn of the century. He made such fiendish demands on the instrumentalists that the Emperor, who was the orchestra's patron, called him in and said, "Don't you think you're overdoing it?" Mahler answered, "Your Majesty, my demands are nothing compared to the demands the musicians now make on me because they play so much better."

You want performers to put on the pressure. You want them to ask: Why can't we do more? Why can't we do better?

BUILDING THE TEAM

The more successful an organization becomes, the more it needs to build teams. In fact, non-profit organizations most often fumble and lose their way despite great ability at the top and a dedicated staff because they fail to build teams. A brilliant man or woman at the top working with "helpers" functions only to a very limited extent; the organization outgrows what one person can do. Yet teams do not develop themselves—they require systematic hard work.

To build a successful team, you don't start out with people—you start out with the job. You ask: What are we trying to do? Then, what are the key activities? I watched from the sidelines while a very effective team management built the fastest-growing labor union in this country. The fellow at the top was an egomaniac, but he knew how to ask the right questions: What are we trying to do? The answer was, We are trying to build a labor organization of poorly paid, unskilled workers doing menial work in hospitals. Next question: What are the key activities needed to achieve our results? Then, and only then, do you ask, What does each of the dozen people at the top have by way of strength? How do the activities and skills match? Within a year, they had a team that would go ahead and expand its union from a membership of fifty thousand to a membership of close to a million in less than ten years. Everybody on the team knew what to do. And, just as important, everyone knew what each one of the other people was going to do. You identify individual strengths, then you match the strengths with key activities. And position your players to take action.

A common mistake is to believe that because individuals are all on the same team, they all think alike and act alike. Not so. The purpose of a team is to make the strengths of each person effective, and his or her weaknesses irrelevant. One manages *individuals* on a team. The focus is to look at the performance and the strengths of individuals combined in a joint effort.

PERSONAL EFFECTIVENESS ON THE JOB

Once the right match is made, there are two keys to a person's effectiveness in an organization. One is that the person understands clearly what he or she is going to do and doesn't ride off in all directions. The other is that each person takes the responsibility for thinking through what he or she needs to do the job. That done, the person goes to all the others on whom he depends—the superior, the associates, the subordinates—and says, "This is what *you* are doing that helps me. This is what *you* are doing that hampers me. And what do *I* do that helps you? What do *I* do that hampers you?" That's 80 percent of working effectively. (But don't write memos. Go and ask!)

The individual who goes through these steps every six months will find that most obstacles disappear. An executive's first responsibility is to enable people who want to do the job, who are paid for doing the job, who supposedly have the skills to do the job, to be able to do it. Give them the tools they need, the information they need, and get rid of the things that trip them up, hamper them, slow them down. But the only way to find out what those things are is to ask. Don't guess—go and ask.

As an organization grows, the non-profit executive must also encourage people at all levels to ask themselves: What does our top management really have to know? I call that educating the boss. It fosters cohesion by forcing individuals to look beyond the scope of their own efforts, their own departments, their own regional needs.

THE TOUGH DECISION

There is an old saying that every soldier has a right to competent command. An effective non-profit executive owes it to the organization to have a competent staff wherever performance is needed. To allow non-performers to stay on means letting down both the organization and the cause.

One common problem is the person who has been in the same job twenty-two years and clearly finds no more stimulus left in it. While a first-class artist never gets tired of his or her work, the rest of us usually get bored if we do the same thing for too many years. The solution is "repotting"—to put the person in a different environment. Again and again I've seen a controller leave a business and move into a hospital. He or she does exactly the same work; only the language is slightly different. But suddenly that person is twenty years younger. The middle-aged burnout, too, usually needs only new demands to come back to life.

A tougher problem is the conflict non-profit executives often face between the need to ensure competence and the need for compassion. But the executives who agonize over this decision do worse than those who say, "We made a mistake. I cut. It's going to hurt, but I cut." It's usually cleaner, faster, and less painful.

THE SUCCESSION DECISION

The most critical people decision, and the one that is hardest to undo, is the succession to the top. It's the most difficult because every such decision is really a gamble. The only test of performance in the top position is performance in the top position—and there is very little preparation for it. Every time we elect a president in the United States we pray that Providence hasn't forgotten America. And that's just as true of lesser top jobs.

What not to do is fairly simple. You don't want a carbon copy of the outgoing CEO. If the outgoing CEO says, "Joe [or Mary] is just like me thirty years ago," that's a carbon copy—and carbon copies are always weak. Be a little leery, too, of the faithful assistant who for eighteen years has been at the boss's side anticipating his or her every wish, but has never made a decision alone. By and large, people who are willing and able to make decisions don't stay in that role very long. Stay away, too, from the anointed crown price. Nine times out of ten that's a person who has managed to avoid ever being put in a position where performance is essential, measured, and where he or she might make a mistake. They are media events rather than performers.

What are the positive ways to handle the succession decision? Look at the assignment. In this community college, in this hospital, in this Boy Scout Council, in this church, what is going to be the biggest challenge over the next few years? Then look at the people and their performance. Match the need against proven performance.

In the end, what decides whether a non-profit institution succeeds or fails is its ability to attract and to hold committed people. Once it loses that capacity, it's downhill for the institution, and this is terribly hard to reverse.

Are we attracting the right people? Are we holding them? Are we developing them? I think you want to ask all three questions about the organization's people decisions. Are we attracting people we are willing to entrust this organization to? Are we developing them so that they are going to be better than we are? Are we holding them, inspiring them, recognizing them? Are we, in other words, building for tomorrow in our people decisions, or are we settling for the convenient and the easy today?

2

The Key Relationships

One of the most basic differences between non-profit organizations and businesses is that the typical non-profit has so many more relationships that are vitally important. In all but the very biggest businesses, the key relationships are few—employees, customers, and owners, and that's it. Every non-profit organization has a multitude of constituencies and has to work out the relationship with each of them.

Begin with the board. In most businesses, boards take little interest in the company until there is a crisis. In the typical non-profit organization, on the other hand, the board is deeply committed. Indeed, non-profit executives and staff often complain that the board is too much concerned with managing, and that the line between board function and management is constantly being violated. They complain that the board "meddles."

To be effective, a non-profit needs a strong board, but a board that does the board's work. The board not only helps think through the institution's mission, it is the guardian of that mission, and makes sure the organization lives up to its basic commitment. The board has the job of making sure the non-profit has competent management—and the *right* management. The board's role is to appraise the performance of the organization. And in a crisis, the board members may have to be firefighters.

The board is also the premier fund-raising organ of a non-profit organization—one important role it does not have in the for-profit business. If a board doesn't actively lead in fund development, it's very hard to get the funds the organization needs. Personally, I like a board that not only gets other people to give money but whose members put the organization first and foremost on their own list of donations.

A board that understands its real obligations and sets goals for its own performance won't meddle. But if you leave the board's role open and undefined, you'll get one that interferes with details and yet doesn't do its job.

Wherever I've seen a non-profit institution with a strong board that gives the right kind of leadership, it represented very hard work on the part of the chief executive officer—not only to bring the right people onto the board but to meld them into a team and point them in the right direction. In my experience, the chief executive officer is the conscience of the board. That may explain why the strong, effective boards I've seen are almost all boards where members come on through a nominating process. I very rarely have seen a truly strong board in co-ops, for instance, where boards are elected by the membership. There the chairperson has no say about who sits on the board, nor has the CEO. Then you get boards which may represent this or that segment of the membership, but they don't represent the organization, at least in my experience. Problems are likely to arise on these boards, such as the troublemakers who abuse the board to create a political platform for themselves or just to hear themselves talk.

Over the door to the non-profit's boardroom there should be an inscription in big letters that says: *Membership on this board is not power, it is responsibility.* Some non-profit board members still feel that they are there for the same reason they used to go on hospital boards in the old days—recognition by the community—rather than because of a commitment to service. Board membership means responsibility not just to the organization but to the board itself, to the staff, and to the institution's mission.

Which brings up a very controversial question, that of an age limit. For a good many older people, membership on the board of a service organization is the last activity they have. They have retired from everything else. So they hold on to it. All my life I've been opposed to age limits. But when it comes to boards, I have reluctantly come around to the idea that it is best to limit membership to two terms of, say, three years each. After that you go off the board. Three years later you may come back on again. But at age seventy-two or so, you go off the board and *stay* off the board.

Another common problem is the badly split board. Every time an issue comes up, the board members fight out their basic policy rift. This is much more likely to happen in non-profit institutions precisely because the mission is, and should be, so important. In my experience, the role of the board then becomes both more important and more controversial. At that point, teamwork between chairperson and chief executive officer becomes absolutely vital.

TWO-WAY RELATIONSHIPS

Only two-way relationships work. Every organization wants stars and needs stars. But in a good opera performance, the star is not separate from the cast. The cast supports the star and then, as the great singer delivers an outstanding performance, the supporting cast is suddenly lifted out of its mediocrity. Everybody suddenly has a new dimension. That's the payoff of an effective two-way relationship.

An effective non-profit executive starts building this two-way relationship with the staff, with the board, with the community, with donors, with volunteers, and with alumni by asking: "What do you have to tell me?" Not, "This is what I am telling you." That question brings problems out in the open. And the funny thing is that most of the problems that bother people so much turn out to be non-problems when you bring them out in the open. A friend of mine calls them "pebbles in the shoe" for which you don't need an orthopedic surgeon. The two-way relationship converts a lot of problems into pebbles in the shoe.

The true test of a relationship is not that it can solve problems but that it can function *despite* problems. Problems don't become irrelevant. But they don't get in the way of what's important.

RELATIONS WITH THE COMMUNITY

The Visiting Nurses, the Cancer Society, the community college, and any number of other non-profit institutions serve one specific community interest. Each has to maintain relations with governmental agencies, with all the other institutions in the community, and with the community's people in general. This is not a matter of public relations (though you'd better have good PR). It requires that the service organization *lives* its mission. That is why volunteers are so important. They live in the community and they exemplify the institution's mission. Effective non-profits train their volunteers to represent them in the community. They also should make it easy for volunteers to report back any questions the community has about the work of the institution.

I know an area where there are three competing hospitals. Everyone in the community is full of praise for one of the hospitals, which on any objective evaluation is probably the poorest of the three. What does that hospital do that makes it so visible in the community? Two weeks after a patient is discharged, somebody from the hospital calls up and says, "Mrs. Smith, I'm calling you on behalf of Memorial Hospital to find out how you are." If Mrs. Smith reports she is not doing well, that recovery is slow, the hospital calls again three weeks later. At the end of the year she gets a calendar from the hospital saying, We hope we don't have to see you again, but we still care about you—something sentimental. Everybody knows that this is pure routine. And yet the hospital tells the community what the community *needs* to hear from a hospital: We haven't forgotten you.

Far too few service organizations even know who their "alumni"—ex-patients or graduates—are. That, I think, is probably the one area where each non-profit manager can easily improve the institution's community standing. Results can be achieved with little effort.

3

From Volunteers to Unpaid Staff

Interview with Father Leo Bartel*

PETER DRUCKER: Am I right, Father Leo, that your diocese has greatly increased the size and scope of its services even though you have far fewer priests and Sisters than you used to have? How did you accomplish this Miracle of the Loaves and Fishes?

LEO BARTEL: In part by hiring lay people to do work that used to be done by priests and Sisters. But primarily we are expanding through volunteers doing a larger and larger share of the diocese's work. We now have at least two thousand volunteers working for the diocese, most of them, of course, women.

PETER DRUCKER: That's news? I thought you in the Catholic Church always had a great many women volunteers.

LEO BARTEL: Of course. But the volunteers of the past were "helpers." Our volunteers now are "colleagues." In fact, we shouldn't even talk of "volunteers" anymore; they are really "unpaid staff." A good many of these people are now in leadership positions in the Church and in Church work.

*Father Leo Bartel is Vicar for Social Ministry of the Catholic Diocese of Rockford, Illinois.

PETER DRUCKER: So the same woman who forty years ago would be arranging the lilies for Easter is now teaching, or taking care of preschoolers, or running the admissions office in the hospital, or chairing the parish council?

LEO BARTEL: Exactly. And this seems to me is a real transformation.

PETER DRUCKER: And how did you manage it?

LEO BARTEL: The need became evident, and on a parish level in particular. The need, I think, shot up first when the Sisters were no longer able to staff the religious education programs for youngsters. Sometimes there are no Sisters now at all even to lead the religious education programs. And so we began to ask lay people. At first as an expedient. Later on we learned that it is not only a good thing, but in many respects it strengthens and encourages and enriches the lives of the volunteers, of the folks who are coming to help. So the pastor might invite people to come into the religious education program—and then we try to provide as well as we can the sort of training and support that will enable our unpaid staff people to do what they're setting out to do. Saturday workshops, in-service days with the religious education directors, and so on.

We have an event we call the Rockford Area Religious Ed Conference, which is becoming pretty well known in our area. Our lay teachers come to Rockford for three or four days and participate in workshops there. Besides that we now have—sponsored by the diocese—the Lay Ministry Formation Program, which takes especially qualified and especially interested lay people from the parish level and eventually certifies them after training as leadership people who are available then in the parish.

PETER DRUCKER: How much training and what kind of training do you give?

LEO BARTEL: The formal training in the Lay Leadership Program runs over a two-year period. We have seven courses, which range from scripture to communications to evangelization to theology. This program is intended to take people who have shown ability and give them the kind of training that will make them effective, give them a sense of being qualified.

PETER DRUCKER: That sounds like a very rigorous program, not too different from what used to be the program for the first vows.

LEO BARTEL: As a matter of fact, it's very, very similar.

PETER DRUCKER: How many people do you have in that program?

LEO BARTEL: A hundred to one hundred and twenty, at this time.

PETER DRUCKER: What's the dropout rate?

LEO BARTEL: So far it's been very, very small.

PETER DRUCKER: That's a remarkable achievement. This is a very demanding program, and not only in time.

LEO BARTEL: Peter, I think if there's one thing that I find exciting about being a priest right now, and about being a parish priest in particular, it's that God seems to be touching more and more people with this invitation to ministry, and they have a thirst, a yearning for this kind of preparation.

PETER DRUCKER: Then volunteers have a common vision and they have dedication and you provide the training. But still, Father Leo, how do you maintain quality control?

LEO BARTEL: It seems to me, Peter, that the quality control is maintained because of the common vision. These people are truly dedicated. And we can depend on their goodwill.

PETER DRUCKER: That's no substitute for knowing what to do as marriage counselor.

LEO BARTEL: The fact is though that if people are properly motivated—and these people are deeply motivated—developing competence becomes part of their very need. My biggest difficulty in asking people to serve is that they are painfully aware of their lack of experience and lack of preparation. If we can provide them with that, they're eager to learn.

PETER DRUCKER: So, what you are saying is rather than lack of competence, the thing you have to worry about is lack of self-confidence. You have to encourage and cheer them on, praise them, help them, be there to support them. But the rest they do.

LEO BARTEL: Besides that, Peter, we hold them to high standards. We have high expectations for them. I believe firmly that people will tend to live up to the expectations that others have for them. And I try, as best I can, to hold high expectations for the people around me, and in many cases they seem to find this a compliment. They seem to be honored that I would expect them to do well. And they come back looking for ways to improve, eager for opportunities to become more and more competent.

PETER DRUCKER: In the hospitals or the schools which are under your direct jurisdiction, how do you do that? Do you sit down with them and work out standards, or benchmarks?

LEO BARTEL: We use many of the common management tools. Then we spend time together trying to develop and articulate a vision, articulate ideals, and articulate priorities that we can all share. We are very careful to develop opportunities for individuals to share their difficulties as well as their triumphs with each other. We give them opportunities to deepen in themselves and in each other the sense of how important the things are that they are doing.

PETER DRUCKER: So you don't treat them as "volunteers" but as staff members. The only difference is that they are part time and they are not being paid. But when it comes to performance, performance is performance.

LEO BARTEL: Absolutely. Competence is competence.

PETER DRUCKER: Father Leo, what do you do with someone who just is not competent, no matter how hard he or she tries?

LEO BARTEL: Sometimes I have to go to somebody and say, "Mary, I'm sorry this is not coming off the way that you want it to and I know that you're not satisfied. Can we talk about this?"

PETER DRUCKER: Perhaps for most of them it's a relief. The person knows perfectly well that he or she doesn't do the job, but isn't able to face up to it and come to you and say, "Get me out of there." They feel they are letting the Church down.

LEO BARTEL: No question about it.

PETER DRUCKER: And you come and say, "We have looked at you and what you are good at is over here. It's not where you are." You are really helping the person. But very few executives understand that. Most bosses shirk it.

LEO BARTEL: The relief is frequently there. But it takes a lot of courage to say a discouraging word. It takes a lot of courage to face somebody and offer them an alternative, because it's easy for the supervisor to think that person is going to feel devalued. Yet in many cases it's a great relief to the person.

PETER DRUCKER: May I switch, Father, and ask you now whether you have any question regarding the management and development of people? Perhaps not just volunteers, people in general?

LEO BARTEL: There are many questions, I'm sure, but two occur to me very quickly. One is a matter of inspiration. How does one excite and motivate folks who are apathetic? The other is the one of organization: When people are going to participate in boards or councils, how do we get them to do the sort of paperwork, the sort of planning work, that they really must do in order to be effective in their roles on councils or boards?

PETER DRUCKER: I'm delighted that you're asking questions I can answer. It doesn't happen too often. Those two questions are very closely connected. But if you talk about inspiring the laity, all I can tell you is that it's the wrong question. We have learned that one inspires the leaders. I once helped run a rapidly growing professional school in which I had to hire very young people who had never taught. And I had to throw them in and run large classes of advanced and demanding students. Every one of these green teachers came to me and asked, "What do I do?" I said, "Make sure you don't lose the top 10 percent of the class. If you lose those, you've lost everybody. But if the top 10 percent are excited and learn, then the average student will learn." The bottom—for those, one prays. But if you don't get the top inspired, you have lost everybody. So, you are doing the right thing with that volunteer program of yours. You are creating a community of achievers. You are doing very much what St. Paul tried to get through to those dense numbskulls, the Corinthians, again and again and again.

Now, your second question, the planning of the parish councils you've formed, and the parish school boards. Make sure that you are not abdicating your responsibility as the CEO, the chief executive Officer, the parish priest. Boards have to be given their work plan. They need leadership, they need to know what the parish needs from them. They need to be told, "You are our associate. We need this from you. We need planning. We don't need you to clean the floors, we need you to plan." And I, the parish priest, or I, the Diocesan Vicar, need somebody to whom I can talk freely. The parish needs you to plan ahead the money-raising campaign for next year. It needs you to think through whether we should rebuild the school through sixth grade. And do we now tackle again that junior high school or high school we were forced to abandon fifteen years ago? We still have the building but that's all we have. That's what the board is there for.

And then the board is there for specific assignments. One says, "Louise, are you willing to go up to Rockford and sit down with Father Bartel and discuss this problem we have here where we need a little more money to do this or do that?" This is an assignment.

Typically, non-profit organizations don't use that tremendous asset, the board—its excitement, the willingness, the commitment. As a result, the board then meddles and becomes petty. And it's up to the CEO right away to say, "This is the work of the board, the work of the advisory council." If the CEO doesn't do it, boards will splinter themselves.

LEO BARTEL: That's a big help, Peter. That's exactly the sort of thing that I've been concerned about. We're looking at revitalizing some boards. The factors that were involved in that were very important for me right now, so that's a big help.

PETER DRUCKER: And as to that apathy, don't forget that Jesus picked only twelve Apostles. If he had picked sixty, he couldn't have done it. He had a hard enough time with those twelve, always saying to them, "Don't you understand?" And it took a long time even for those handpicked, very exceptional young people. So, one works with the leaders because there is a rule in human affairs that the gap between the leaders and the average is a constant. You see it in sports, you see it in music, you see it in almost any area. The job of the leader is to set high standards, by example. For what one person does, another human being can always do again.

LEO BARTEL: Once there is a precedent, then it can be done again. The four-minute mile comes to mind.

PETER DRUCKER: I go back to the days where the five-minute mile was considered beyond human capacity. I was in high school then, and all of us knew "Five minutes for the mile, the good Lord hasn't created the human body to run any faster." Then one fine day, in the early twenties, a Finn broke it, and six weeks later all of us had whittled six seconds off our mile. That's the way it goes.

Let's switch topics: Is there any one guiding principle you have in managing a heterogeneous group of volunteers, and a rapidly growing one?

LEO BARTEL: Peter, I try more than anything to keep central my conviction of the dignity of each person. Each person has the same dignity as a child of God, and it seems to me most important to meet each of these persons freshly each day with a sense of how important they are to God, and therefore, how important they should be to me.

There's another aspect of this that has to do with the task. A person is never going to have a sense of his own, her own, dignity unless they are able to fulfill the expectation of completing the tasks and discharging the responsibilities that they take on. As their supervisor, it seems to me that I have to, on the one hand, keep in mind most importantly that they are children of God, but also, that they probably will not be successful in understanding this and perceiving this for themselves unless they are able to do very well at the responsibilities they are assigned to. So it seems to me extremely significant that I have to do anything I can or provide anything that we can as the Church, so that these colleagues of mine will be successful in what they set out to do.

PETER DRUCKER: One of my mentors and teachers during World War II said to me: "Young man, if you ever grow up, you will learn that one needs both St. Paul and St. James." One needs faith *and* works, is that what you are saying?

LEO BARTEL: Absolutely.

PETER DRUCKER: This is one of the profound insights, and one learns that in managing people. But you also told me something about how you make operational that belief in the dignity of each human being as a creature of the Lord. You see your job as helping people achieve.

LEO BARTEL: The person who is constantly falling short of his own expectations, the person who is constantly thwarted in the things that he or she takes on, will never get to the point where he or she has a sense of their dignity, a sense of their own worth. If they fail, I've failed. And their success is my success.

PETER DRUCKER: Yes, there is no greater achievement than to help a few people get the right things done. That's perhaps the only satisfactory definition of being a leader.

4

The Effective Board

Interview with Dr. David Hubbard*

PETER DRUCKER: You have built an outstandingly effective board at Fuller Theological Seminary, David. How do you see the func tion of the board in the non-profit institution?

DAVID HUBBARD: We need to think of the management of schools, hospitals, churches, and of non-profit institutions altogether as a partnership between the board and the professional staff. I use a side-by-side organization chart, with the board of trustees in one column and the faculty in another column, and the president's office and the various members of the administrative team in between. All three are centers of power, and centers of authority. My task is to promote understanding and fellowship and relationships between those centers, and keep them running parallel so they don't pull apart or collide.

PETER DRUCKER: What specifically does this imply for the board's role?

DAVID HUBBARD: A board needs to know that it owns the organization. But it owns an organization not for its own sake—as a board—but for the sake of the mission which that organization is

*Reverend David Allan Hubbard is president of Fuller Theological Seminary in Pasadena, California.

to perform. Board members don't own it as though they were stockholders voting blocks of stock; they own it because they care. I would say there's often a wrong understanding on the parts of boards of what that ownership means. They actually own it in partnership because, in a sense, the organization belongs just as much to others.

PETER DRUCKER: And how do you create this partnership?

DAVID HUBBARD: It starts, of course, with the way the mission of the institution is stated. And that mission itself needs to be stated with sufficient breadth to allow for flexibility. The mission needs to be welcoming of change. Then you need people who are open to that mission. If you find that the board becomes inflexible, you have to look for ways of renewing the board with fresh appointments, with two or three key people who change the balance of power on the board. The more power is concentrated in a few people on a board, the more likely the situation will turn unhealthy.

Our board at Fuller does not have a rotation system where members go off the board automatically every three or five years. Many organizations do, and there is much to commend it. We have chosen to take a tougher line—to evaluate performance when a board member's term is up. If we think that a trustee has performed effectively in terms of attendance and participation and stewardship, and understanding and so forth, then we will ask that trustee whether he or she is willing to serve further. If not, then we thank them for their service, and tell them that we will be replacing them with someone else, and that the newcomer will bring perhaps another quality that we need. We are fair on our performance evaluation. But for board members who do perform, we like long, continuous service. In higher education, continuity is important. Learning how the institution works takes literally years. But also, the older people are, the more free they are to disperse their capital and to plan their estate and to give away portions of their estate.

PETER DRUCKER: Who makes that decision to appoint or not to reappoint?

DAVID HUBBARD: The Trustee Affairs Committee makes that. That's a group of half a dozen senior trustees. They make that decision—usually on the recommendation of the CEO.

PETER DRUCKER: Do you work closely with that committee?

DAVID HUBBARD: Very closely.

PETER DRUCKER: And you mentioned another function of the board, which is the money-raising function. Do you look upon your board as the leader in raising money?

DAVID HUBBARD: We do. In fact, I might just tick off how I would see the functions of a board member and we can talk about each specifically. Board members are governors. When they sit around the table and vote their "I so move," they govern the institution. Board members are sponsors, and here we get to their role in giving money and raising money. They are ambassadors—interpreting the mission of the institution, defending it when it's under pressure, representing it in their constituencies and communities. Finally, they are consultants; almost every trustee will have some professional skill which would be expensive if you had to buy it. I can call certain trustees and ask a legal question or an administrative question or an educational question and get an almost instant reaction. Governor, sponsor, ambassador, and consultant would be the four major roles.

Now, when it comes to the sponsor role, when we recruit trustees we say to them: "We expect you to give proportionate to your means, and in your giving to assign a high priority to our institution. Your local church and perhaps one other organization can be as important to you as Fuller, but we don't want Fuller to be any lower than third, and we would prefer Fuller to be second behind the commitment to the local church." I also will talk to them about including Fuller as part of their estate, because ultimately with trustees you not only want year-by-year contributions. You want to participate in some form or another, through trusts or annuities or wills, in the final distribution of their wealth.

PETER DRUCKER: So you want a very active board member. You have regular trustee meetings. He or she serves on committees. You want them to be available to you for consultation in their area of expertise. And you look to them as the leaders in money-raising. How many days a year, actually, does this take?

DAVID HUBBARD: It averages eight to ten days a year, including board meetings, perhaps a special committee assignment, extra reading, and then some duty of entertaining on behalf of the seminary or serving it in some way in their own community. We also take them periodically on study tours. We found that very effective. There's an investment of time by the trustee, but also, I need to underscore, an investment of the CEO and the staff in the servicing of those trustees.

PETER DRUCKER: So you consider making the board effective and keeping it effective among the CEO's priority tasks?

DAVID HUBBARD: I think a CEO has two primary areas of service. I have to care for the vice-presidents, whom I supervise, and who have no other boss but me. And I have to care for the trustees, who have no other direct and immediate and ongoing contact with the institution beside me and what my office staff does. In fact, I have one assistant whose key priority, aside from managing my own schedule, is to service the board of trustees.

PETER DRUCKER: How do you balance board involvement with the possibility of board meddling? For example, a board member who gets to know the head of a department and begins to meddle. How do you deal with that?

DAVID HUBBARD: You try to take that innovative energy and channel it into the process. You try to get the board member to talk about his or her concern in the board meeting. Our board meets three times a year; in every session there will be at least one hour when board members can form their own agenda on the spot. We call it open forum. The board member can bring up the subject he wanted to discuss with the department head at that time and if the board wants to take a look at it, they can kick it back to the administration for review. They can put it on the agenda of the appropriate board committee. It is then channeled to the normal process.

PETER DRUCKER: Again and again I hear the professional heads of a non-profit institution say, "Let's not go to the board with this. It's much too controversial." You've heard that, haven't you? I've always felt that one of the things CEOs have to learn is that a subject belongs at the board level precisely because a subject *is* controversial—and the sooner the better. Am I right?

DAVID HUBBARD: You are right on target, Peter. (A) We share bad news first. (B) We tell bad news at 110 percent and good news at 90 percent in order to compensate for our tendency to cheat, almost unconsciously, because we want to tell the board all the good news and we want to minimize the bad news. And that is exactly wrong. Ducking controversy or minimizing difficulty, snowing people with reports that are not realistic either about the quality of the program or about the financial stability or whatever, that's terrible leadership.

PETER DRUCKER: The last thing a non-profit executive should want is for their board to read in the paper something about the institution they run that they didn't learn before. The executive loses all credibility.

DAVID HUBBARD: It's the old principle of no surprises for the boss. Keeping a board well informed is hard work. It takes time, it takes communication—time on the phone and sending out a notice or a report in a preliminary way, or mobilizing the staff and saying to each vice-president, You call these seven or eight trustees and you tell them this. And do it today, on the phone, get the message through. Then the calls come back and the correspondence starts. That's all labor-intensive. But we have no choice but to do it.

PETER DRUCKER: And how do you handle the situation where you need a board to change its position, for instance, to adopt a change in an old, outmoded, but cherished policy?

DAVID HUBBARD: We always try to work for a win situation. We try to help the trustees to change their minds or to expand their vision without feeling that they are letting go of their own cherished goals. Those things are best done one on one. Presentations to an entire board without a lot of spadework, when feelings are strong and attitudes are entrenched, is very difficult. A board can con itself into unity and take a unanimous adversary stance to a proposal unless there is a lot of preliminary conversation on a one-on-one basis to develop advocacy for the idea within the board. The board has to have its own internal advocates.

The style I have developed over the years is to use a point person, the committee chairperson, for instance, for the kind of change that I am proposing. I can then be very passive in a board meeting because a lot of time has gone into personal cultivation and education and orientation of the key board members, and they run interference and carry the ball.

PETER DRUCKER: How do you do this and avoid the board splitting into factions? You can't talk to everybody about everything, can you?

DAVID HUBBARD: No. You have to talk to the people who would be viewed as the point person on a particular issue. If it's an academic issue, you ordinarily would work through the chair of that committee. Ditto with facilities or development. Then there are untitled board leaders. There are patriarchs and matriarchs who have the esteem of the board because of their wisdom or their financial contribution and their loyalty, their stature. You try to work with them, too. And you look particularly for pockets of opposition and work with them. You know on any given issue that someone will help you lead it, but someone else will be very sensitive to that. You have to work both sides and prepare the person, who may not, at first glance, look like a supporter, for the fact that the subject is coming up. You say, "You may not like it or support it. I'm not asking you to, but let me explain in a little detail why I think we need to do it." You give that person perfect freedom to oppose it. But you have given the grace or the courtesy of anticipating their objections.

If someone loses in a board vote, I make it my aim at the first possible break to go to the person who lost and thank him or her for the courage to express a contrary opinion. As president, my task is not only to shape the majority so we move in a positive direction, but also to comfort, support, and encourage the minority. I would sum it up under a heading, something like integrity. What you're doing is just *not* a matter of strategy. At heart it's respecting the dignity of trusteeship, of directorship.

PETER DRUCKER: It's hard to do what you have just told us with boards which, unlike yours, are outside boards by intent—the elected school board, the city council. There the CEO, the school superintendent or the city manager, tends to see board members as enemies or adversaries. The less we tell them, the better. He tries to play politics and then he loses.

But in my experience, even on such boards, your way is the only way to operate, especially on school boards, which have become very political. The school board I knew the best had the very difficult problem of desegregating a community that had been gerrymandered to keep black kids out of white schools. It was a highly explosive issue. And the superintendent succeeded because he had respect for the integrity and function of the board. It wasn't always easy because the board was badly split. But he started out by asking: What do we all have in common? We all are dedicated to enabling the kids to learn. Let's start out with this. Over five tough, bitter years, he succeeded. A neighboring community had a very much smarter superintendent who felt the board could never agree on anything, so it was his job to prevent doing harm by not telling them anything, by being clever. He lasted only eighteen months. That community is still embroiled in fights they haven't resolved.

DAVID HUBBARD: You know, they are called trustees because they are trusted. But trustees also need to be *trustors* to function well—they have to trust the CEO. Anything the CEO does to lose credibility with the trustees, even when adversity is hot, when the quarrel is sparking, anything that person does to lose credibility will ultimately make it impossible for that person to function.

There is nobody clever enough to outsmart a board over any length of time and succeed. Even if you succeed short term, the whole thing turns to chalk because you don't have that sense of integrity. In your writing, Peter, you've stressed so much that the process is essential to the quality of the product. And the process of trusteeship is one of the central processes in organizational life. The process of leadership with the board is as central to the successful outcome—hospital care or relief—as any other single task.

PETER DRUCKER: Let me try to sum up the most important things I've heard.

The most important thing I heard, you didn't say, you implied—that it is to the benefit of an institution to have a strong board. The tendency of so many CEOs is to try to have a board that won't do any harm because it won't do anything. It is the wrong tendency. You depend on the board, and therefore you can be more effective with a strong board, a committed board, an energetic board, than with a rubber stamp. The rubber stamp will, in the end, not stamp at all when you most need it.

The second reason is that to get this strong board, the non-profit executive has to do a lot of very hard work. Good boards don't descend from Heaven. It requires continuing work to find the right people, and to train them. They come in knowing what you expect of them, and they have very tough expectations in terms of time and money and work and responsibility. You take a great deal of time to keep the board informed but also to have a two-way flow of information.

And building relationships with the board is a crucial, central part of the task of the CEO.

Does this sum up, more or less, what you have been telling us, David?

DAVID HUBBARD: That's an excellent summary, Peter, and I would just underscore the value of all of this to an organization. An organization hasn't come anywhere near its full potential unless it sees the building of a great and effective board as part of the ministry of that organization.

5

Summary:
The Action Implications

In no area are the differences greater between businesses and non-profit institutions than in managing people and relationships. Although successful business executives have learned that workers are not entirely motivated by paychecks or promotions—they need more—the need is even greater in non-profit institutions. Even paid staff in these organizations need achievement, the satisfaction of service, or they become alienated and even hostile. After all, what's the point of working in a non-profit institution if one doesn't make a clear contribution?

Furthermore, there are people working in non-profit organizations that businesses have no experience dealing with. They are called "volunteers," though that no longer is quite the right word. They are different from the paid workers in a non-profit only in that they are not paid. There is less and less difference between the work they do and that done by the paid workers—in many cases it is now identical—and the volunteers are becoming increasingly important to non-profit organizations. Not only is the number of volunteers increasing. They are taking on more and more leadership functions. This trend is likely to continue as we have many more older people in our society who are capable of working physically and mentally and are also eager to stay active, to stay involved, to contribute. Thus, non-profit institutions will be serving their specific missions, but they also will increasingly be the organizations through which we make citizenship operational and effective.

Altogether, the non-profit executive deals with a greater variety of stakeholders and constituencies than the average business executive. The non-profit institution's relationship with its donors, for instance, is not known to business enterprises. A company's shareholders and customers have completely different expectations from donors. The non-profit board also plays a very different role from the company board. It is more active and, at the same time, more of a resource if managed properly—and more of a problem if *not* managed properly. The problems can be most pronounced when the board is not selected by the institution itself but is elected—as co-op boards and most school boards are elected—by outside constituencies which may be critics of the institution.

Because of the complexity of relationships for the non-profit executive, it is important to understand and apply what we know about the management of people and the management of relationships. And we know a good deal.

People require clear assignments. That's true of volunteers; that's true of the board; that's true of the employed staff. They need to know what the institution expects of them. But the responsibility for developing the work plan, the job description, and the assignment should always be on the people who do the work.

The non-profit executive must work both with employed staff and with volunteers so that they can think through their contribution, spell it out clearly, and evolve by joint discussion a specific work plan, with specific goals and specific deadlines. The less control you have over people in the old-fashioned sense by fear, disciplining them, demoting them, or not promoting them, the more important it is that they have a clear assignment for which they themselves take responsibility.

The non-profit must be information-based. It must be structured around information that flows up from the individuals doing the work to the people at the top—the ones who are, in the end, accountable—and around information flowing down, too. This flow of information is essential because a non-profit organization has to be a learning organization. Emphasis in managing people should always be on performance. But, especially for a non-profit, it must also be compassionate. People work in non-profits because they believe in the cause. They owe performance, and the executive owes them compassion. People given a second chance usually come through. If people try, give them a second chance. If people try again and they still do not perform, they may be in the wrong spot. Then one asks: Where should he or she be? Perhaps in another position in the organization—or perhaps elsewhere, in another organization. But, if a person doesn't try at all, encourage him or her as soon as possible to go to work for the competition.

A recurring problem for non-profit organizations such as churches, hospitals, and the Scouts are the people who volunteer because they are profoundly lonely. When it works, these volunteers can do a great deal for the organization—and the organization, by giving them a community, gives even more back to them. But sometimes these people for psychological or emotional reasons simply cannot work with other people; they are noisy, intrusive, abrasive, rude. Non-profit executives have to face up to that reality. Perhaps there is a job, in some corner, which they can do. But if there isn't, they must be asked to leave. The alternative is that the executive, and all those who have to work with the person, lose capacity to contribute.

The non-profit board is both the tool of the non-profit chief executive and the chief executive's conscience. For this relationship to prosper, the chief executive must develop a clear work plan for the board. A non-profit executive can—and must—manage even a board that is elected by outside (sometimes critical) forces and that cannot be dismissed by the professional executive. But, to be productive, the board must be informed. The worst thing a chief executive can do is try to hide things from the board, play little games, focus on finding a friend or two on the board and ignore building an overall relationship. That's always a temptation. But the executive who yields to it can be guaranteed to be out on his or her ear within a year or two.

Everyone in the non-profit institution, whether chief executive or volunteer foot soldier, needs first to think through his or her own assignment. What should this institution hold me accountable for? The next responsibility is to make sure that the people with whom you work and on whom you depend understand what you intend to concentrate on, and what you should be held accountable for.

Next are the learning and teaching responsibilities: What do I have to learn? What does this organization have to learn? Not in five years—but now, over the next few months. If you are an executive in a non-profit institution, make sure you sit down next week with your key people and say, "I am not here to tell you anything. I am here to listen. What do I need to know about you and your aspirations for yourselves—and for this organization of ours? Where do you see opportunities that we don't seem to be taking advantage of? Where do you see threats? What are we doing well? What are we doing badly? What improvements do we have to make?"

Make sure to listen—but also make sure to take action on what you hear and learn.

Ask every one of the people who report to you or with whom you work: "What am I doing that helps you with your work? What am I doing that hampers you?" Act on what they tell you. If the complaint is, for instance, that you don't give information unless asked, make sure that the required information goes out every Friday, or whenever. If they say they don't know how they are doing, build feedback into your system. They have *their* jobs to do, and the non-profit executive's job is to enable them to do it, successfully and satisfactorily. What you may need most, along with your associates, is clear information about the results of your organization's work. We go out and solicit money by talking needs. That's fine. But both donors and people who work for a non-profit inevitably ask—What are the results? No executive should respond with generalities.

The effective non-profit executive finally takes responsibility for making it *easy* for people to do their work, *easy* to have results, *easy* to enjoy their work. It's not enough for them, or for you, that they serve a good cause. The executive's job is to make sure that they get *results*.

PART FIVE

Developing Yourself

*as a person, as an executive,
as a leader*

1

You Are Responsible

The first priority for the non-profit executive's own development is to strive for excellence. That brings satisfaction and self-respect. Workmanship counts, not just because it makes such a difference in the quality of the job done but because it makes such a difference in the person doing the job. Without craftsmanship, there is neither a good job, nor self-respect, nor personal growth. Many years ago I asked the best dentist I ever had, "What do you want to be remembered for?" And he answered, "When they have you on the autopsy slab, I want them to say that fellow really had a first-rate dentist!"

How different that attitude is from the person who does the job to get by, who hopes that nobody will notice.

Self-development is very deeply meshed in with the mission of the organization, with commitment and belief that the work done in this church or this school matters. You cannot allow the lack of resources, of money, of people, and of time (always the scarcest) to overwhelm you and become the excuse for shoddy work. Then you begin to blame the world—"they" won't let me do a good job. And that's the first step down a steep, slippery slope. Paying serious attention to self-development—your own and that of everyone in the organization—is not a luxury for non-profit executives. Most people don't continue to work for a non-profit organization if they don't share, at least in part, the vision of the organization. Volunteers, particularly, who don't get a great deal out of working for the organization aren't going to be around very long. They don't get a check, so they have to get even more out of the organization's work. In fact, you don't want people who stay on with the organization just because that's what they've always done but who don't believe in it anymore. And in a well-run, results-oriented organization, you should be making such demands on people for time and work that it's unlikely too many with that jaded outlook would stay on. You want constructive discontent. That may mean that many of the best of the paid staff or volunteers come home exhausted after a big meeting, complaining loudly about how stupid everybody is and how they don't do things that are obvious, and then respond, "But it's so important!" if someone asks why they stay on.

The key to building an organization with such a spirit is organizing the work so everyone feels essential to a goal they believe in. One of the church people I work with has a clear goal that in this church of twelve thousand members, there are no parishioners. There are only paid and unpaid ministers—everyone is put to work at that level. That's a goal; not yet an accomplishment. Nevertheless, working toward that goal, from fifty to a hundred people a year are added to the force taking on church responsibilities. By now the church has almost no paid staff. Instead of the usual paid, ordained, youth minister, this church has six unpaid and unordained individuals who, together, do the one full-time job. And each of these volunteers sits down twice a year and writes a letter to himself or herself (a copy to the pastor) answering the questions: "What have I learned? What difference to my own life has my work with kids at the church been making?" The pastor has no difficulty attracting volunteers. In fact, his problem is a waiting list.

TO MAKE A DIFFERENCE

From the chief executive of a non-profit on down through the ranks of paid staff and volunteers, the person with the most responsibility for an individual's development is the person himself—not the boss. Everyone involved must be encouraged to ask themselves: What should I focus on so that, if it's done really well, it will make a difference both to the organization and to me? A hospital floor nurse, for example, under terrific pressure of time and money, with doctors demanding more and the paperwork out of control, looks at the thirty-two orthopedic patients and says, "*They* are my job. All the rest, basically, are impediments. What can I do to concentrate on that job? Maybe it is something procedural. Can we change the way we deliver our services to enable me to be a better nurse?"

You can only make *yourself* effective—not anyone else. Your first responsibility to the non-profit organization for which you work is to make sure you get the most out of yourself—for yourself. You can work only with what you have.

Creating a record of performance is the only thing that will encourage people to trust you and support you. Complaining about stupid bosses, a stupid board, and subordinates who sabotage you, won't create that record. It's *your* job and *your* responsibility to talk to those on whom you depend, and who depend on you, to find out in a systematic way what helps, what hinders, and what needs to be changed.

All the people I've known who have grown review once or twice a year what they have actually done, which part of that work makes sense, and what they should concentrate on. I've been in consulting for almost fifty years now and I've learned to sit down with myself for two weeks in August and review my work over the past year. First, where have I made an impact? Where do my clients need me—not just *want* me but need me? Then, where have I been wasting their time and mine? Where should I concentrate next year so as not only to give my best but also to get the most out of it? I'm not saying that I always follow my own plan. Very often something comes in over the transom and I forget all my good intentions. But so far as I have become a better and more effective consultant and have gotten more and more personally out of consulting, it's been because of this practice of focusing on where I can really make a difference.

Only by focusing effort in a thoughtful and organized way can a non-profit executive move to the big step in self-development: how to move beyond simply aligning his or her vision with that of the organization to making that personal vision productive. Executives who make a really special contribution enable the organization to see itself as having a bigger mission than the one it has inherited. To expand both the organization and the people within it in this way, the top executive must ask the key questions of himself—the questions I ask myself each August. Indeed, each member of the staff must do it, and each volunteer. And the senior people must sit down regularly with each other and consider the questions together.

The form for this kind of exchange can be quite flexible. In fact, one of the best examples I've ever heard of was improvised by Bruno Walter, the great conductor, much loved by the musicians he led. At the end of each season, Walter wrote a letter to each member of the orchestra something like this: "My dear [First Trumpet], you taught me quite a bit when we rehearsed the Haydn symphony by the way you handled that difficult passage. But what have you learned this season as a result of our working together?" Probably half the musicians simply sent back a perfunctory postcard. But the other half sat down and wrote serious letters: "I now suddenly understand what I, as a twentieth-century trumpeter, have to do to sound like an eighteenth-century trumpeter in the Haydn symphony." Playing in Bruno Walter's orchestra became a constant developmental challenge for his musicians.

The critical factor for achieving this kind of success is accountability—holding *yourself* accountable. Everything else flows from that. When you are president or vice-president of the university or the bank, the important thing is not that you have rank, but that you have *responsibility*. To be accountable, you must take the job seriously enough to recognize: I've got to grow up to the job. Sometimes that means acquiring skills. Even harder, you may find that the skills you worked so hard to acquire over the years no longer apply: you spent ten years learning all about computers, but now you have to learn to work with people. By putting accountability first, you build the commitment to mobilize your own resources. You ask: What do I have to learn and what do I have to do to make a difference? A wise person I worked with many years ago said to me, "For good performance, we give a raise. But we promote only those people who leave behind a bigger job than the one they initially took on."

Self-development seems to me to mean both acquiring more capacity and also more weight as a person altogether. By focusing on accountability, people take a bigger view of themselves. That's not vanity, not pride, but it is self-respect and self-confidence. It's something that, once gained, can't be taken away from a person. It's outside of me but also inside of me.

SETTING AN EXAMPLE

In all human affairs there is a constant relationship between the performance and achievement of the leaders, the record setters, and the rest. In human affairs, we stand on the shoulders of our predecessors. The leader sets the vision, the standard. But he or she is not the only one. If one member of an organization does a markedly better job, others challenge themselves.

Leadership is not characterized by stars on your shoulder; an executive leads by example. And the greatest example is precisely the dedication to the mission of the organization as a means of making yourself bigger—respecting yourself more.

2

What Do You Want to Be Remembered For?

To develop yourself, you have to be doing the right work in the right kind of organization. The basic question is: "Where do I belong as a person?" This requires understanding what kind of work environment you need to do your best. When young people come out of school, they know very little about themselves. They do not know whether they work best in a big organization or a small one. They rarely know whether they like working with people or working alone, whether they prosper in a situation of uncertainty or not, whether they need the pressure of deadlines to perform efficiently, whether they make decisions quickly or need to sleep on them. The first job is a lottery. The chances of being in the right place are not very good. It takes a few years to find out where you belong and to begin self-placement.

We all tend to take temperament and personality for granted. But it's very important to take them seriously and to understand them clearly because they're not too subject to change by training. People who have to understand a decision completely before they can act don't belong on a battlefield: when the right flank suddenly caves in, an officer may have eight seconds to decide whether to fight or retreat. The kind of person who likes to reflect on decisions might force himself to decide—but is very likely to make the wrong decisions.

If the thoughtful answer to the question "Where do I belong?" is that you don't belong where you currently work, the next question is why? Is it because you can't accept the values of the organization? Is the organization corrupt? That will certainly damage you, because you become cynical and contemptuous of yourself if you find yourself in a situation where the values are incompatible with your own. Or you might find yourself working for a boss who corrupts because he's a politician or because she's concerned only with her career. Or—most tricky of all—a boss whom you admire fails in the crucial duty of a boss: to support, foster, and promote capable subordinates.

The right decision is to quit if you are in the wrong place, if it is basically corrupt, or if your performance is not being recognized. Promotion itself is not the important thing. What is important is to be eligible, to be equally considered. If you are not in such a situation, you will all too soon begin to accept a second-rate opinion of yourself.

"REPOTTING" YOURSELF

Sometimes a change—a big change or a small change—is essential in order to stimulate yourself again. Recognizing this need will grow in importance as people live for many more years than they used to and are active so much longer. A great many volunteers, for instance, move on to another organization after ten or twelve years of working for one non-profit. The usual need they feel is to change the routine. An unexpressed need may be that they no longer are learning. Be aware of that touchstone yourself, because when you stop learning in a job, you begin to shrink.

The switch doesn't have to be to something far afield. Richard Schubert, for instance, for many years president of the American Red Cross, came up as a labor lawyer and human resources manager in private industry. In his forties, he switched to government and then back to private industry—and then to the Red Cross. He is so effective precisely because he has worked with a wide variety of different people in quite different work cultures.

When you begin to fall into a pleasant routine, it is time to force yourself to do something different. "Burnout," much of the time, is a cop-out for being bored. Nothing creates more fatigue than having to force yourself to go to work in the morning when you don't give a damn.

Perhaps all that is needed is a small shift—the school principal who accepts a few invitations to visit other school districts and talk over problems with other principals and teachers. The other possibility is to take on a volunteer job with another organization. That might seem impossible to non-profit executives who are already working sixty to seventy hours a week; but three hours a week spent in an entirely different activity might do the trick. Precisely because you are overworked, you need the extra—and different—stimulus to put different parts of yourself to work, both physically and mentally. The Girl Scouts now have many more volunteers than they ever had in their history because they discovered that busy women working as lawyers and accountants and bank officers also need the challenge of working hard in an entirely different environment.

Most work is doing the same thing again and again. The excitement is not the job—it is the result. Nose to the grindstone, eyes on the hills. If you allow a job to bore you, you have stopped working for results. Your eyes, as well as your nose, are then on the grindstone.

To build learning into your work, and keep it there, build in organized feedback from results to expectations. Identify the key activities in your work—perhaps even in your life. When you engage in such activities, write down what you *expect* to happen. Nine months or a year later, compare your expectations to what actually happened. From that you will learn what you do well, what skills and knowledge you need to acquire, what bad habits you have (which might be the most important discovery). Or you may find out, as I did, that you stopped too soon in your push for results. I soon realized that I'm terribly impatient. You may also realize that, again and again, your best intentions do not produce results because you don't listen—the most common bad habit.

You're certainly not limited to learning only from your own

activities. Look at the people in your own organization, your own environment, your own set of acquaintances. What do they do really well—and how do they do it? In other words, look for successes. What does Joe do that seems so hard for the rest of us to do? Then try to do it yourself. It's up to you to manage your job and your career. To understand where you best belong. To make high demands on yourself by way of contribution to the work of the organization itself. To practice what I call preventive hygiene so as not to allow yourself to become bored. To build in challenges.

DOING THE RIGHT THINGS WELL

Most of us who work in organizations work at a surprisingly low yield of effectiveness. I've been working with executives for close to fifty years and most of them work hard and know a great deal. But fully effective ones are rare. The difference between the performers and non-performers is not a matter of talent. Effectiveness is more a matter of habits of behavior, and of a few elementary rules. But the human race is not too good at it yet because organizations are pretty recent inventions. The rules for effectiveness are different in an organization from what they were in the one-man craft shop. In solo work, the job organizes the performer; in an organization, the performer organizes the job.

The first step toward effectiveness is to decide what are the right things to do. Efficiency, which is doing things right, is irrelevant until you work on the right things. Decide your priorities, where to concentrate. Work with your own strengths. The road to effectiveness is not to mimic the behavior of the successful boss you so admire, or to follow the program of a book (even mine). You can only be effective by working with your own set of strengths, a set of strengths that are as distinctive as your fingerprints. Your job is to make effective what you have—not what you don't have.

You identify strengths by performance. There is some correlation between what you and I *like* to do and what we do well. There is a strong correlation between what we hate to do and what we do poorly simply because we try to get it out of the way as fast as possible, with minimum effort and postpone, postpone, postpone working on it at all. Albert Einstein said he would have given everything, including the Nobel Prize, for the ability to play the fiddle well enough to play in a symphony orchestra. But he simply didn't have the coordination between his two arms and hands that are the prerequisite for being a master string instrument player. He loved playing—he practiced four hours a day and enjoyed it. But it wasn't his strength. He always said he hated doing math. He was only a genius at math.

Strengths are not skills, they are capacities. The question is not, can you read, but are you a reader or a listener, for instance? This particular characteristic is almost as strongly determined as handedness. Franklin D. Roosevelt and Harry S Truman were listeners. Roosevelt rarely read anything; he had it read to him. Eisenhower was a reader but didn't know it. When he was Commander-in-Chief in Europe his press conferences were widely admired. His aide insisted that journalists hand in their questions—written—up to a few minutes before the conference. Ike read them and his responses were superb. Then he became president, following Roosevelt and Truman, who had set the style of taking questions from the press from the floor (as listeners, they were good at it). Ike, however, performed poorly; the press disliked him because they said he never answered the question. His eyes glazed over. He didn't even really hear the questions.

People have become more understanding in recent years of how strengths vary from person to person—that there are morning people, or perceptive people, or conceptual people. What many people do not know about their strengths and weaknesses, however, is whether they are comfortable with other people or have to learn how to work with them. Too many think they are wonderful with people because they talk well. They don't realize that being wonderful with people means *listening* well.

SELF-RENEWAL

Expect the job to provide stimulus only if you work on your own self-renewal, only if you create the excitement, the challenge, the transformation that makes an old job enriching over and over again. Seeing both yourself and the task in a new dimension can sometimes expand this capacity. There is an old story about the great clarinetist in an orchestra who was asked by the conductor to sit in the audience and listen to the orchestra play. For the first time, he heard music. He wasn't simply playing the clarinet expertly, he was making music. That's self-renewal. Not doing anything differently but giving it new meaning.

The most effective road to self-renewal is to look for the unexpected success and run with it. Most people brush the evidence of success aside because they are so problem-focused. The reports executives usually work with are also problem-focused—with a front page that summarizes all the areas in which the organization underperformed during the past period. Non-profit executives should make the first page show the areas where the organization *overperformed* against plan or budget, because that is where the first signs of unexpected success begin to appear. The first few times you will brush it aside: Leave me alone, I'm busy solving problems. Eventually, though, a suspicion may begin to surface that some of the problems would work themselves out if we paid more attention to the things that were working exceptionally well. I know a very able woman who runs a small community service agency. She began to notice that her Visiting Nurses were putting in steadily increasing claims for overtime. First, like all of us, she tried to control the increase. She met with the nurses, asked them why their overtime bills were climbing, and discovered that they were treating more people after 6:00 P.M., when they came home from work. As a result of improved medical care, the caseload was shifting from invalids and shut-ins to people who functioned but who needed help with services such as insulin therapy, physical rehabilitation, injections. Now she is in a new field. She is a missionary to meet this new need—and she has become a newly vigorous and effective person.

The three most common forcing tools for sustaining the process of self-renewal are teaching, going outside the organization, and serving down in the ranks. When an individual is asked to explain to a group of colleagues how she did something that worked very well, she learns, and so do the listeners. Spending time doing volunteer work in another organization also opens up alternatives. And one of the oldest techniques for keeping executives alive to the realities of implementing an organization's mission is for them to work once or twice a year at the level where service is delivered to the organization's clients. One well-trained medical bureaucrat I know was forced by a strike or some sudden epidemic years ago to work as a floor nurse for a week. Suddenly he was down where the heartbreaks and the successes were played out. It forced him to learn and, as he admitted to me, "It forced me to be honest with myself." Now the hospital's rule (and it is one of the finest hospitals I know) is that he and all his administrators spend one week a year working on the floor with the people who take care of the patients.

All the individuals who have the greatest ability for self-renewal focus their efforts. In a way, they are self-centered, and see the whole world as nourishment for their growth.

WHAT DO YOU WANT TO BE REMEMBERED FOR?

When I was thirteen, I had an inspiring teacher of religion, who one day went right through the class of boys asking each one, "What do you want to be remembered for?" None of us, of course, could give an answer. So, he chuckled and said, "I didn't expect you to be able to answer it. But if you still can't answer it by the time you're fifty, you will have wasted your life." We eventually had a sixtieth reunion of that high school class. Most of us were still alive, but we hadn't seen each other since we graduated, and so the talk at first was a little stilted. Then one of the fellows asked,

"Do you remember Father Pfliegler and that question?" We all remembered it. And each one said it had made all the difference to him, although they didn't really understand that until they were in their forties.

At twenty-five, some of us began trying to answer it and, by and large, answered it foolishly. Joseph Schumpeter, one of the greatest economists of this century, claimed at twenty-five that he wanted to be remembered as the best horseman in Europe, the greatest lover in Europe, and as a great economist. By age sixty, just before he died, he was asked the question again. He no longer talked of horsemanship and he no longer talked of women. He said he wanted to be remembered as the man who had given an early warning of the dangers of inflation. That is what he is remembered for—and it's worthwhile being remembered for. Asking that question changed him, even though the answer he gave at twenty-five was singularly stupid, even for a young man of twenty-five.

I'm always asking that question: What do you want to be remembered for? It is a question that induces you to renew yourself, because it pushes you to see yourself as a different person—the person you can *become.* If you are fortunate, someone with the moral authority of a Father Pfliegler will ask you that question early enough in your life so that you will continue to ask it as you go through life.

3

Non-Profits:
The Second Career

Interview with Robert Buford *

PETER DRUCKER: Tell me, Bob, when you decided to add a major non-profit institution, Leadership Network, to your activities and to be the chief executive of it in addition to running your own business, you were in your mid-forties. What did you have to learn to make that transition?

ROBERT BUFORD: The critical thing for me to learn was how to reallocate my own sense of identity from how well I do in business—basically a life of accumulation—to one of service, where service is the primary driving force in life.

PETER DRUCKER: Is that a change in values or a change in behavior or both?

ROBERT BUFORD: I hold the same values I've had all along. But I had to make a major change in proportions and behavior.

*Robert Buford is chairman and CEO of Buford Television, Inc., in Tyler, Texas. He has founded two non-profit institutions, Leadership Network and the Peter F. Drucker Foundation for Non-Profit Management.

PETER DRUCKER: I take it, while you have been very successful in business, you never saw money, even in business, as "the" goal. It's a measurement rather than a goal?

ROBERT BUFORD: Clearly so. But as a score-keeping mechanism, it was important to me and easy to see. I find now that I've undertaken this second career, that the score-keeping mechanism changes, and I need to be very conscious of that. You can choose the game you're in but not the rules of the game. As I have chosen a different game to play as a primary source of my own activity and identity, I've had to be very conscious of changes in the rules of that game. It has required for me a real sense of clarity about mission and goals and about what comes first. But there comes a time in everyone's life when one has to decide what the critical concerns are and what the subordinate concerns are.

PETER DRUCKER: You consider that the critical decision in developing oneself?

ROBERT BUFORD: It's critical to know who your master is. And I think it's critical to update that understanding periodically. I think I am a different person in terms of my desires and how I want to allocate my time, talent, and treasure in my mid-forties than I was in my twenties.

PETER DRUCKER: Has your behavior had to change a great deal, or do you do the same things but to a different purpose and to a different drummer?

ROBERT BUFORD: The latter, I think. I find that what I do for my company is very similar to what I do for Leadership Network. In both cases, I have to be clear about what the vision is so that other people can function successfully and can function as a team. In both cases, I have to encourage and support other people in their work and make the work of either the business or Leadership Network clearly *their* work. And in both cases, I have to maintain a critical set of relationships that teaches me what's going on in those two worlds.

PETER DRUCKER: Priorities might be quite different, though?

ROBERT BUFORD: Leadership Network is that which is exciting to me now. Though I'm still in business, my business is now subordinated. In my twenties, I subordinated my desire to be in the ministry.

PETER DRUCKER: Did you find it very difficult to make that change?

ROBERT BUFORD: I didn't find it difficult. I found it rather like a change of season. It just seemed to me in my mid-forties that it was time to get around to things that were eternal and of great significance and importance. In doing so, I found that I had to make a great many changes in my business career.

PETER DRUCKER: What made you realize that the time had come? Was it just success that enabled you to change, or did you kind of wake up one fine day and say, it's time that I looked at myself?

ROBERT BUFORD: I think, first of all, I accumulated enough "score" to feel comfortable that I'd finished one game. Secondly, a series of experiences have taught me that I am what St. Paul calls a "citizen of eternity." It was simply clear to me that it was time to get on with those concerns.

PETER DRUCKER: So, nothing sudden?

ROBERT BUFORD: Perhaps the difference is that I am now willing to listen to the calling that was there all along. And I'm perhaps better equipped by the experiences of the last twenty years to serve in that calling.

I find I use the same entrepreneurial skills that I've had all along. But I use them for a different purpose and in a different cause. I find that it's very important as you're making these kinds of changes to have a little self-knowledge. And I think my experience of these twenty years has taught me that where I belong is to be an entrepreneur functioning with a team.

PETER DRUCKER: Self-knowledge is as important as task knowledge. And if you are skill-focused rather than task-focused, you miss a turn, so to speak. You keep going down the old road but, all of a sudden, it leads nowhere. *Start on the outside* is what you are saying. Start with: What is the purpose? Who is the master? Then you use the same tools—but you build a different edifice.

ROBERT BUFORD: I think the two questions are the ones which you've taught in your books, and they're enduring and important questions: Who is the customer? And what does the customer consider value? In Leadership Network, I have a different set of customers than I have in my business, and I have to be very sensitive to their values.

PETER DRUCKER: You've had significant achievements in both of your careers. Is there any particular experience that helped you either to do the right things or avoid doing the wrong ones?

ROBERT BUFORD: Perhaps two experiences that came early in my life. My mother gave me a great deal of responsibility early in life and a great deal of freedom to fail. The second thing that was important to me is that I got caught off base a couple of times when I was quite young. For the rest of my life I've assumed that anything I did in violation of the rules, I would get caught doing. So, I've made it a rule that I'm simply not going to take shortcuts and cheat, because I assume I'll get caught. And I find that's good discipline.

PETER DRUCKER: Can you remember any one person in your own company or in your own community who made you realize who really you are and who you might become? For instance, I've heard you talk a great deal about how much you gave, but also how much you got from the Young Presidents Organization. Was that one of the important relationships in your life?

ROBERT BUFORD: The Young Presidents Organization has been important in my life because it's given me a window into the real worlds of other executives. I have chosen to live all my life in a town with a population of seventy-five thousand because it seems to me to be a sane environment to function from, and a caring and warm environment. But it is a small town. The Young Presidents Organization has provided me with access to sophisticated and successful people whom I would otherwise have been unlikely to meet.

PETER DRUCKER: That's why it's so important, I think, for people who work in an organization to have an outside interest, *to meet people* and not just become totally absorbed in their own small world. And all worlds are small worlds. That's particularly important for people in non-profit organizations because their work is so much more absorbing than it is in a business. When you say to a business executive, you're working hard from nine to five, make sure you have some other interest—be a Scout Master, well, that gets a resonance. But when you say to a pastor, perhaps you should go on the board of the local hospital, he says, I'm too busy. He becomes a victim of his own organization. One of the most successful—and busy—non-profit executives I know sits on several company boards. She says that gives her a window on a different world—that she learns from doing that.

Let me ask you what important advice you have on self-development for people in non-profit service organizations? You have seen more of them than almost anybody I know, worked with more of them through your pastoral churches and the service organization executives you work with in Leadership Network. What would be the important advice?

ROBERT BUFORD: On either the business side or the non-profit side, stay in touch with your constituency, or you run the risk that they will change and you won't. You'll be left a prisoner of your own tradition, a prisoner of the insiders in an organization and their desires, and will miss the role of a service organization, which is to serve.

PETER DRUCKER: I'm reminded that Gustav Mahler told his orchestra members they should sit in the audience at least twice a year so that they know what music sounds like to the listener. A great pastor I knew years ago made it his habit to take off about four or five Sundays a year, go to other churches, and sit in the congregation. Is that what you are telling me is important?

ROBERT BUFORD: A great pastor I know summers in the country and goes to small local churches all summer. Another pastor I know makes it his practice to go to the offices of his members on a frequent and disciplined basis to meet them on their turf.

PETER DRUCKER: The best hospital administrators I know have themselves admitted once a year as a patient, go through the admission routine, and then spend a day just to see not only how their organization works but what it is like to be a patient.

So that's one of the important development things. Any other?

ROBERT BUFORD: It's very important that the leader, and, for that matter, the whole leadership team, stay in touch with the seasonal changes within themselves. We all have different experiences and levels of intensity in our mid-forties than we had in our mid-thirties. And we will be entirely different in our mid-fifties when, perhaps, we're bored with our current careers, where we have achieved virtuosity and mastery in things which we used to think very challenging, but which are now yesterday's work.

4

The Woman Executive in the Non-Profit Institution

Interview with Roxanne Spitzer-Lehmann *

PETER DRUCKER: Roxanne, what did the people who first promoted you from a nurse to a manager see in you that made them promote you?

ROXANNE SPITZER-LEHMANN: Organization skills, communications ability, and a great concern for the people I cared for as patients.

PETER DRUCKER: Can you identify where some of those traits came from?

ROXANNE SPITZER-LEHMANN: I was fortunate to have several mentors. I think nursing education has also played a great role in terms of developing the ability to prioritize, to determine how and when to do something. I think what's going to happen in the

*Roxanne Spitzer-Lehmann is corporate vice-president of St. Joseph Health System, a chain of non-profit hospitals headquartered in Orange, California. She is the author of *Nursing Productivity* (Chicago, 1986).

health-care sector, particularly hospitals, is that more nurses will be moving ahead because of that organization ability, because of that ability to prioritize, because of communications skill and the technological knowledge that comes with it.

PETER DRUCKER: What role did your mentors play, Roxanne, in developing these organizational and human skills and in making you aware of their importance?

ROXANNE SPITZER-LEHMANN: I tend to be impatient. And they've helped me look at the data before making a decision. Helped me understand that my basic reaction to problems and/or situations was probably good, but I had to slow up prior to implementing or determining a course of action. Certainly, they've taught me patience. They've also allowed me to make mistakes as well, and I think that's an important factor.

PETER DRUCKER: Any of them ever point out what you do well?

ROXANNE SPITZER-LEHMANN: There was a lot of positive reinforcement.

PETER DRUCKER: Now let me switch to something totally different, Roxanne. Are you the only woman in the top management of the hospital chain today?

ROXANNE SPITZER-LEHMANN: Yes, I'm the only corporate officer who's a woman.

PETER DRUCKER: And how many women are there in top management of major hospitals other than the Catholic Orders?

ROXANNE SPITZER-LEHMANN: Not many, but I think it's increasing. At the present time more are moving into chief operating officer and chief executive officer positions. But certainly relative to an industry that has a very high percentage of females in it, it is very low. Hospitals are very traditional; they are modeled very much on the military. But I think necessity is the mother of invention. And as the need for greater productivity, greater flexibility in roles, and ability to organize becomes imperative in this competitive marketplace, more women will be assuming those roles.

PETER DRUCKER: What advice would you give to women moving into positions of leadership in an institution in which women were very much subordinates, owing absolute obedience to the all-powerful physician who always was a man?

ROXANNE SPITZER-LEHMANN: Any advice I'd give to an executive would probably not be limited to gender. I think that women probably have to do it a little better, and a little harder. But, in fact, the greatest attribute a woman can have going into any organization, and health-care particularly, is to play as a team member. Not to be isolationist, not to be territorial. To be willing to give up in order to have the organization move. To help others give up departments, give up responsibilities; look at matrix organizations as opportunities, not as a loss of power; look at the development of others.

It's been very interesting to my colleagues and myself (and I don't believe I'm alone in this) that as more females enter the medical staff, it has been more difficult for them to work with the other females in the organization than it is for the male physician. Maybe these women physicians are having such difficulty in making it in the male-dominated medical world that they need to be a little bit more aggressive and not quite as supportive to their female colleagues. I think that's a major mistake—for any woman to play the role of queen bee. That is pushing herself away from other women and not working with them to develop them. Of course, women did not usually learn how to play football on a team or baseball, and when one becomes an executive, learning how to play football or baseball with the guys is a real key to success.

PETER DRUCKER: You work pretty closely with a very powerful, very proud board. Did they find it awkward to accept a woman at first? Especially the women on the board?

ROXANNE SPITZER-LEHMANN: My board, like any other hospital board, is predominantly male-dominated. It's only been the last several years that women have been on the board, and there is only one woman now on the executive committee of the board. The women board members have been highly supportive. Those women are generally competent, well-developed business women in their own right, are very comfortable with themselves, and don't need to achieve at the expense of someone else. The women are not a problem at all on the board.

The men are very interesting, depending upon age groups. The older age group certainly had some difficulty accepting a woman in a corporate position. The younger age group is used to working with women, I think, out in the real world. There's a strong sense of paternalism in hospitals. On one hand, it's very protective of me as their only woman vice-president. On the other hand, they make it somewhat clear that they don't really consider me chief executive officer material. That's not universal, but we do talk about it a little.

PETER DRUCKER: Can you think of a specific example of a time when you felt you had broken through these barriers?

ROXANNE SPITZER-LEHMANN: When I did a presentation to the board about a financial program, instead of just reporting on patient care and aspects of clinical delivery services and patient satisfaction and quality-assurance kinds of things. The board suddenly realized that I knew a lot about the profit and loss statement. In fact, I'm about to do another report to the board on my Home Care Department, which is highly profitable. As they see that I'm responsible for the financial aspects as well as the delivery of services, I'm watching a breakthrough occur.

PETER DRUCKER: How did you acquire the skills necessary for you to do that?

ROXANNE SPITZER-LEHMANN: First, from a mentor in my very early years as a director of nursing. I was fortunate to have a mentor from the university who insisted that I learn up front what man-hours per patient per day meant, and how to determine salaries. So, I was always a little bit ahead, I think, of the market in terms of that. Then, of course, being responsible teaches you. My budget now is about $75 million. One learns very rapidly how to look at bottom lines and how to make sure expenses do not exceed revenue, although that's quite difficult today. And, of course, pursuing my doctoral degree in Executive Management at Claremont has been tremendously helpful by crystallizing the details. I had no great problems with overall bottom lines, but I've become almost as astute as our financial department, I think, which makes them a little nervous.

PETER DRUCKER: What about the people skills? A nurse is aware of people and their needs. But she is not really aware of working in an organization. How did you acquire people skills when you moved into the director of nursing position and suddenly worked with sixty, seventy, two hundred other nurses and patients, and had to coordinate nursing with other departments of that New York hospital in which you started out? Did you have to learn the skills, or did they just come naturally?

ROXANNE SPITZER-LEHMANN: I think some people skills come naturally. The ability to coordinate and communicate is something that one learns somewhat through trial and error, somewhat through humility by being willing to listen and to learn when one has made an error in communication. One learns to say, "I'm sorry, I didn't mean it that way." I think that's a real factor.

I think I've always had a vision about what I thought patient care should be and how I thought it should be delivered. I've never had a great problem communicating my vision and then moving toward it. And I've been fortunate in how people buy into that vision with me. It's easy to work with people when you have something in common that makes some sense, that's goal-oriented. So, I think people skills are very much based upon communicating a common goal. And then, of course, you learn over time how many errors you make when you didn't communicate correctly.

PETER DRUCKER: So, you would say the first thing is that vision—which is probably the reason why you went into nursing in the first place, or at least why you stayed in nursing—that vision is really the basis?

ROXANNE SPITZER-LEHMANN: I believe so. I think I also had a cause about women, since nursing is predominantly a female profession. I graduated in the sixties when women were not in very powerful positions, so I had kind of a cause for nursing.

PETER DRUCKER: So you came in with a vision and a cause and the desire to communicate it—really the desire to be a leader. And nobody ever said in those early years, "Roxanne, don't be pushy"?

ROXANNE SPITZER-LEHMANN: Oh, they still say that. And I *was* very pushy. I don't know how many times my bosses and my colleagues have said to me: "Girl, you are aggressive!" But when one really believes in something, it's very hard not to be aggressive. How can anyone argue with "We're not delivering patient care in a way that is best for the patient"? The patient should determine how his or her body should be serviced. That shouldn't be determined by procedures that hospitals design. I came out believing that in the very early years.

PETER DRUCKER: Roxanne, you shock me. In forty years of working in the health-care sector, I've heard nothing but people saying, "Don't listen to the patient. We know what's right."

ROXANNE SPITZER-LEHMANN: I don't know how that's possible. I think patients may not have all the knowledge necessary to make decisions. But it's our responsibility to help them gain that knowledge so that they can make informed decisions.

PETER DRUCKER: So you would say that for any institution that is the starting point—What are we really here for?

ROXANNE SPITZER-LEHMANN: If you don't know the mission, you shouldn't be around.

PETER DRUCKER: Roxanne, you're clearly a woman with a mission. I'd be curious to know how you structure your life and your work to make that mission a reality.

ROXANNE SPITZER-LEHMANN: Well, it's very hectic, I can tell you that—having a full-time job, and a teenage daughter, and still going to school. In fact, going to school and working helped me keep my mission very focused. As does a fifteen-year-old daughter who is one tremendous conscience asking, "Why, Mommy, do you do all of these things?"

One is, of course, self-driven, not always just by a mission but by a need to accomplish. If I didn't have the mission, I'd get a much easier job. Or I'd lie on the beach all day in Southern California. That temptation frequently comes to mind—until a situation occurs in which a really focused intervention can improve either the service delivery or the quality of life of my employees, then the temptation to lie on the beach disappears fast and I am glad I have a tough job.

And now we, in the hospital, face more and more of these challenges.

PETER DRUCKER: A little before your time, the hospital was a very simple organization, with doctors and nurses and a few people who cleaned up. Now it's becoming terribly complicated—all those specialties, all those services. And you see your mission as focusing all of them on that common objective, the patients, who should leave the hospital at least no worse than when they came in.

At the end of the year, how do you know whether you have helped make that mission a greater reality? What are the areas of success? What are the areas where you have to do better?

ROXANNE SPITZER-LEHMANN: Well, there are two ways. One is concrete and one is abstract. The concrete is really easy to describe. I keep a pad on the right side of my desk which I add to or modify, perhaps every two weeks, once a month. On one side it lists the major undertakings that I have to do, and on the other side it lists those that are in process, to whom they've been delegated, and what the status is. When they're completed, I just put a single line through them. At the end of the year, I take a look at this and I'm always overwhelmed at how much we have accomplished. We put together an annual report based upon that.

I also use management-by-objectives to some degree. That is a really concrete way of seeing how we've moved forward.

On an abstract level, I certainly take a look at how I do on my academic work toward my doctorate. Every course I pass seems to be a benchmark toward the future.

But other than that I think it is very difficult. I never feel that I've done enough or that I've achieved enough.

PETER DRUCKER: May I switch completely? You talked about your being responsible for a budget of well over $70 million and for the financial performance of quite a few services. Where do you see the most important differences, in your work as an executive and professional, between a business organization and a non-profit service organization?

ROXANNE SPITZER-LEHMANN: In the health-care sector, we have become so like industry in terms of having to be competitive, in terms of having to be bottom-line-oriented, that I see my role little different from anyone working at General Motors, or Xerox, or IBM. I have a product to deliver. I have to deliver it cost-effectively. I have to make sure that the consumers are satisfied. And they shouldn't have to return—though, certainly, if the need arises, you want them to come back to your institution. We're in a business. We have competition around us, especially in Southern California. We have to deliver something better. Something better and at the right price. That's not very different than Procter & Gamble.

PETER DRUCKER: Roxanne, you haven't really talked much about self-development. You have mentioned mentors. You have mentioned that pad of yours in which you put down your tasks and your accomplishments. But you haven't really talked much about developing yourself.

ROXANNE SPITZER-LEHMANN: I think the best self-development is developing others. I'm fortunate enough that people will tell me when I'm wrong, when I come on too strong, and when I don't give them enough time to do their own thinking.

PETER DRUCKER: What are you doing to encourage your associates to grow and develop themselves? What are the things that have been most effective?

ROXANNE SPITZER-LEHMANN: My role is not to give the answers. My role is to facilitate their brainstorming and thinking. And then to pull it together into something that we all go out and implement. My job is to establish the goal and the vision. *Their* job is to figure out how we can do it together. And I believe that allowing, promoting, giving people the time—time is an important element—the skills, the tools, and the environment to do that has enhanced my self-improvement. I've become relatively well known in the industry because my staff has been so creative in what we determine together. If I were to leave tomorrow, I don't think it would make much difference. They would carry on.

PETER DRUCKER: You come from a profession which is known for a high degree of what's frequently called burnout—people feeling the pressures are just too great. There must be days or weeks when you feel that pressure. How do you renew yourself?

ROXANNE SPITZER-LEHMANN: That's a question the whole industry is asking, with the nursing shortage looming as a major catastrophe. Self-renewal comes from feeling good about oneself. The nurses at the bedsides can feel good about themselves if they're given the autonomy and control to do what they do best. My self-renewal comes from being given autonomy, respect, the control to take a project from the beginning to the end without a lot of interference in getting it done.

The best example I can give you of that was when we first opened our outpatient surgery center. Everybody had been diddling with it for years until I finally said, "Would you just let me do it? Would you just let me put together the elements and carry it through?" It was given to me lock, stock, and barrel, and we did it. I got a great deal of self-renewal from that. My other self-renewal comes from a personal life. I like to cook. I love the theater. I love music. I've learned how to ski in the last year and fall down a great deal, which has been great for self-renewal. And I like to travel. That's all self-renewal.

PETER DRUCKER: Well, that's a classical answer to burnout. The way to overcome burnout is to work much harder. And it apparently works for you. I must testify it also works for me. But then you have enough things that are totally different from your work. The theater, falling down on the ski slope, music—you change mental and emotional gears a little bit. I think that's very important.

Let me try to pick out a few major strands. To me, the most telling thing you said is, "If I were to leave tomorrow, I don't think it would make much difference. They would carry on." That's about the proudest boast any executive can make, to have built the team that will perpetuate *my* work, *my* vision, *my* institution.

That, in my experience, really distinguishes the true achiever.

Then you stressed the importance of the mission and of the focus on the desired results: cured patients. And you stressed again and again team building. That is leadership in developing others, which can be the most important key to self-development.

5

Summary:
The Action Implications

The best way for me to start this summary on self-development is to tell you about the man who first made me aware of what that means as a lifelong process. He was a Jewish rabbi whom I first met in the early 1950s on a mountain trail. We became hiking companions for many years because we both spent vacations in the same summer resort and liked hiking. Joshua Abrams had been in law school when World War II broke out, went into the Navy and was badly wounded. In fact, he never fully recovered, and the injuries eventually caused his death thirty-five years later.

He went into a seminary when he came out of the service and, when I first met him, he had just begun to build—from scratch—a synagogue and Jewish community center in a major Midwestern city. Just ten years later it was one of the largest Reformed Jewish synagogues in the country, with four to five thousand members.

So, I was very surprised on a walk one day when he said, "By the way, Peter, I've decided to leave the synagogue and start all over again." I looked at him, clearly without understanding, and he continued, "I don't learn anything anymore." A year later, he told me he had decided to go into youth ministry and take over the chaplainship at a major Midwestern university. This was about 1964–65. Joshua explained: "I'm still young enough so that I understand what troubles the kids and I'm old enough to have experience with most of the things that they are going through. They're going to be in trouble." Sure enough, the student unrest started not too long afterward and my friend was a tower of strength. Through the years I've met people who say, "I understand you know Josh Abrams? He saved my life when I was twenty years old and about to destroy myself by going into drugs . . . or by doing this, that or the other stupid thing."

Then, around 1973–74, Josh surprised me again during one of our walks: "I think I've done all I can do as a university chaplain. I'm no longer young enough to be in tune with the kids. I've been thinking about it and have decided that the need now is for a ministry for old people. That's where the population growth is." He quit the university a year or two later, moved to one of the retirement cities in Arizona, and started all over again building from scratch. By the time he died, his new community of retired people was three to four thousand strong. He looked for people who were lonely, who had lost their spouses, who were sick, and he not only brought them spiritual comfort but helped meet their physical needs as well as he could.

Joshua was the first person who explained something to me that I have, in turn, repeated to many, many people: "You are responsible for allocating your life. Nobody else will do it for you." And the pattern of his life makes clear that when we talk of self-development, we mean two things: developing the person, and developing the skill, competence, and ability to contribute. These are two quite different tasks.

Developing yourself begins by *serving,* by striving toward an idea outside of yourself—not by leading. Leaders are not born, nor are they made—they are self-made.

To do this, a person needs focus. Michael Kami, our leading authority on business strategy today, draws a square on the blackboard and asks: "Tell me what to put in there. Jesus? Or money? I can help you develop a strategy for either one, but you have to decide which is the master."

I do it by asking people what they want to be remembered for—that's "the beginning of adulthood," according to St. Augustine. The answer changes as we mature—as it should. But unless that question is asked, a person works without focus, without direction, and, as a result, does not develop. You start by developing your own strengths, adding skills and putting them to productive work. There is much a boss can do to contribute to this development. But no matter how much a boss drives you—or ignores you—ultimately it is the individual's own responsibility to work on his or her own development.

Developing your strengths does not mean ignoring your weaknesses. On the contrary, one is always conscious of them. But one can only overcome weakness by developing strengths. Don't take shortcuts. You don't have to be a perfectionist but you certainly should refuse to accept your own shoddy work. Above all, workmanship builds your own self-respect as it builds your own competence.

Next, you work on the tasks to be done, the opportunities to be explored. And you start with the task, not with yourself. Achievement comes from matching need and opportunity on the outside with competence and strength on the inside. The two have to meet—and the two have to match.

Effective self-development must proceed along two parallel streams. One is improvement—to do better what you already do reasonably well. The second is change—to do something different. Both are essential. It is a mistake to focus only on change and forget what you already do well. One works constantly on doing a little better, identifying the little step that will make the next step possible. But it is equally foolish to focus only on improvement and forget that the time will inevitably come to do something new and quite different.

Listening for the signal that it is time to change is an essential skill for self-development. Change when you are successful—not when you're in trouble. Look carefully at your daily work, your daily tasks, and ask: "Would I go into this today knowing what I know today? Am I producing results or just relaxing in a comfortable routine, spending effort on something that no longer produces results?"

Self-development becomes self-renewal when you walk a different path, become aware of a different horizon, move toward a different destination. This is a time when outside help, a mentor, can provide useful help. The more achievement-minded and successful you are, the more likely you are to be immersed in the task at hand, immersed, above all, in the urgent. A wise outsider who knows what you are trying to do, who has often been doing similar things, is the one who can ask you: "Does it still make sense? Are you still getting the most out of yourself?"

The means for self-development are not obscure. Many achievers have discovered that teaching is one of the most successful tools. The teacher usually learns far more than the student. Not everybody is in a situation where the opportunity to teach opens up, nor is everyone good at teaching or enjoying it. But everyone has an associated opportunity—the opportunity to help develop others. Everyone who has sat down with subordinates or associates in an honest effort to improve their performance and results understands what a potent tool the process is for self-development.

Probably the best of the nuts and bolts of self-development is the practice of keeping score on yourself. It's also the best lesson in humility, as I can tell you from personal experience. It is always painful for me to see how great the gap is between what I should have done and what I did do. But, slowly, I improve—both in setting goals and in achieving results. Keeping score helps me focus my efforts in areas where I have impact and to slough off projects where nothing is happening, where I'm wasting not only my own resources but also those of my clients or students.

Self-development is neither a philosophy nor good intentions. Self-renewal is not a warm glow. Both are action. You become a bigger person, yes; but, most of all, you become a more effective and committed person. So, I conclude by asking you to ask yourself, what will you *do* tomorrow as a result of reading this book? And what will you *stop doing?*

Index

For information about the audiocassette series, "The Nonprofit Drucker," available on 25 one-hour cassettes with listener's guide, and the Peter F. Drucker Foundation for Nonprofit Management, contact the Drucker Foundation.

The Peter F. Drucker Foundation for Nonprofit Management, founded in 1990, is named for and inspired by the acknowledged father of modern management. By providing educational opportunities and resources, the Foundation furthers its mission, "To lead social sector organizations toward excellence in performance." The Foundation pursues this mission through the presentation of conferences, video teleconferences, the annual Peter F. Drucker Award for Nonprofit Innovation, and the development of management resources and publications.

If you would like more information on the Drucker Foundation and its programs and publications, or if you would like to support its work with a contribution, please contact the Foundation.

The Drucker Foundation
320 Park Avenue, 3rd Floor
New York, NY 10022-6839 USA

Tel: 212-224-1174
Fax: 212-224-2508
Email: info@pfdf.org